Submarines and Underwater Exploration

Bruce LaFontaine

DOVER PUBLICATIONS, INC.
Mineola, New York

Bruce LaFontaine

Bruce LaFontaine is the illustrator and writer of more than twenty-four non-fiction children's books. His published works are sold in bookstores throughout the United States, Canada, and the United Kingdom. Mr. LaFontaine specializes in history, science, transportation, and architectural subjects for the children's middle-reader market (ages 8–12). He lives and works in the Rochester, New York area.

Copyright

Copyright © 1999 by Bruce LaFontaine
All rights reserved under Pan American and International Copyright Conventions.

Published in Canada by General Publishing Company, Ltd., 30 Lesmill Road, Don Mills, Toronto, Ontario.

Bibliographical Note

Submarines and Underwater Exploration is a new work, first published by Dover Publications, Inc., in 1999.

DOVER *Pictorial Archive* SERIES

This book belongs to the Dover Pictorial Archive Series. You may use the designs and illustrations for graphics and crafts applications, free and without special permission, provided that you include no more than four in the same publication or project. (For permission for additional use, please write to Permissions Department, Dover Publications, Inc., 31 East 2nd Street, Mineola, N.Y. 11501.)

However, republication or reproduction of any illustration by any other graphic service, whether it be in a book or in any other design resource, is strictly prohibited.

International Standard Book Number: 0-486-40803-5

Manufactured in the United States of America
Dover Publications, Inc., 31 East 2nd Street, Mineola, N.Y. 11501

INTRODUCTION

The great oceans of our world cover an expansive seventy percent of the earth's surface, so it is natural that human beings have always been irresistibly drawn to the power and mystery of the sea and intrigued by the possibility of its exploration: as a source of food, for cover in warfare, to search for treasure, to observe marine life, and for the sheer adventure of charting a new and unknown frontier.

Ancient artifacts offer evidence that man has attempted to penetrate the depths of the oceans for thousands of years. Stone carvings from the kingdom of Assyria (circa 1100 B.C.) depict divers walking on the sea bottom using primitive breathing apparatus fashioned from inflated goatskins—a primitive version of the aqualung. Attempts at underwater exploration were made by various other early civilizations, including that of Greece, Rome and medieval Europe. Aristotle mentions sponge divers and even a rudimentary form of diving-bell in his writings, and the Roman historian Livy tells of treasure divers. However, these efforts were not ultimately successful for the simple reason that machinery and materials were not advanced enough to facilitate safe underwater activity for prolonged periods of time. In fact, it was not until the scientific and technological innovations of the Industrial Revolution—beginning about 1700—that undersea exploration became a real possibility, and eventually, a reality.

Unlocking the secrets of what lies beneath the sea has followed a two-pronged approach. One led to the development of diving equipment used by individuals for exploration and research; while the other—fueled by the aggressive demands of warfare—led to the evolution of undersea boats, commonly called submarines.

By the end of the eighteenth century, a primitive submarine had been built that was actually used for battle. In 1776, the American colonial navy deployed the wooden submarine, *Turtle*, against the British during the War for Independence from Great Britain. By the mid-nineteenth century, many nations were experimenting with undersea waterships that incorporated numerous features of modern vessels, including ballast tanks and diving planes for submerging, electrically-driven propellers for propulsion, oxygen-generating equipment, and torpedoes for underwater attack.

At the same time that these early submarines were being constructed, safe and efficient "individual" underwater diving gear was becoming a practical reality. The classic "hard hat" helmeted diver began with equipment invented in 1819 by Augustus Siebe, and later perfected by him in 1838. Using Siebe's apparatus, divers were finally able to move about on the seabed for extended periods of time, breathing air pumped to them by a surface ship.

But as with much technological advancement, warfare-driven submarine development moved at a more rapid pace. In 1897, Irish-American engineer John Holland invented the first class of modern naval submarines for widespread use. His prototype, the USS *Holland*, became the model for the turn-of-the-century submarine fleets of several nations. But it was not until World War I that the submarine evolved into an effective weapon; and by the Second World War, it had become a decisive strategic weapon. Multiple submarine "Wolfpacks" operated by the German navy prowled beneath the Atlantic and wreaked terrible destruction on Allied shipping. Finally, as an outgrowth of the atomic bomb used to end World War II, nuclear power was adapted to submarine propulsion, so that in 1955 the United States Navy launched the SSN-571 *Nautilus*, the first of these very advanced nuclear-powered submarine warships.

The undeclared "Cold War" between the United States and the Soviet Union further spurred development of submarines from the 1960s through the conflict's end in the early 1990s. During this period, both nations constructed powerful ballistic missile-launching subs capable of delivering devastating nuclear strikes. These antagonists also built hunter-killer "attack" submarines specifically designed to find and destroy enemy missile boats.

Fortunately, one area of underwater activity during wartime actually led to a major advance in "civilian" research and exploration of the undersea world. In 1943, French naval officer Jacques-Yves Cousteau and engineer Émile Gagnan invented the

SCUBA—or Self-Contained Underwater Breathing Apparatus—diving system. This breathing equipment allows divers to swim unencumbered by an air hose connected to a surface vessel. Impelled by the flexibility and freedom of this innovation, underwater exploration has become widespread.

Advanced technology borrowed from submarine warships has also led to the development of sophisticated undersea research vessels called "submersibles," which are used strictly for exploration, rescue, and salvage. In 1960, the submersible bathyscaphe *Trieste* descended to the deepest part of the Pacific Ocean—an incredible 35,817 feet below the surface. Other research submersibles such as the *Alvin, Aluminaut,* and *Nautile* have greatly expanded our knowledge of the deep ocean areas. Robot submersibles called "ROVs" (Remotely Operated Vehicles) are now used to explore sunken ships and photograph the deep sea bottom.

Of all places on earth, the sea remains our richest untapped resource until we learn how to live and work underwater. And while we currently know a great deal about this vast watery domain that encompasses more than two-thirds of the total surface area of our planet, there is still much to be learned about the deepest reaches of the seas—a frontier that is as mysterious and beckoning as outer space.

Assyrian Diver with Primitive Aqualung, 1100 B.C.–600 B.C.

The earliest record of attempts to explore the briny deep can be found on carved stone tablets from the ancient empire of Assyria, which show divers walking on the sea bottom carrying leather bags filled with air. Attached to the air sacks was a tube that allowed the diver to breathe from them. This primitive attempt at underwater activity was arduous, impractical, and dangerous. It had no mechanism for closing off the diver's nose to prevent taking in water, and the air supply was short and difficult to regulate.

One problem the Assyrians may have attempted to solve was that of a diver's natural buoyancy—i.e., the tendency to float in the water rather than walk upright on the bottom. They may have equipped the diver with a container filled with heavy weights, such as rocks or metal. The diver above is shown utilizing such a system.

Historical cultures of the past were probably motivated to experiment with underwater exploration for several reasons. They no doubt saw the ability to roam the sea bottom as an easy and effective way to hunt marine animals for food. Another was the possibility of using divers in warfare, since attacking enemy ships and coastal fortifications from beneath the ocean was likely seen as a very effective means of gaining advantage over an enemy. This would ultimately prove to be a viable notion, but its practical application during actual warfare would not be accomplished for another several thousand years.

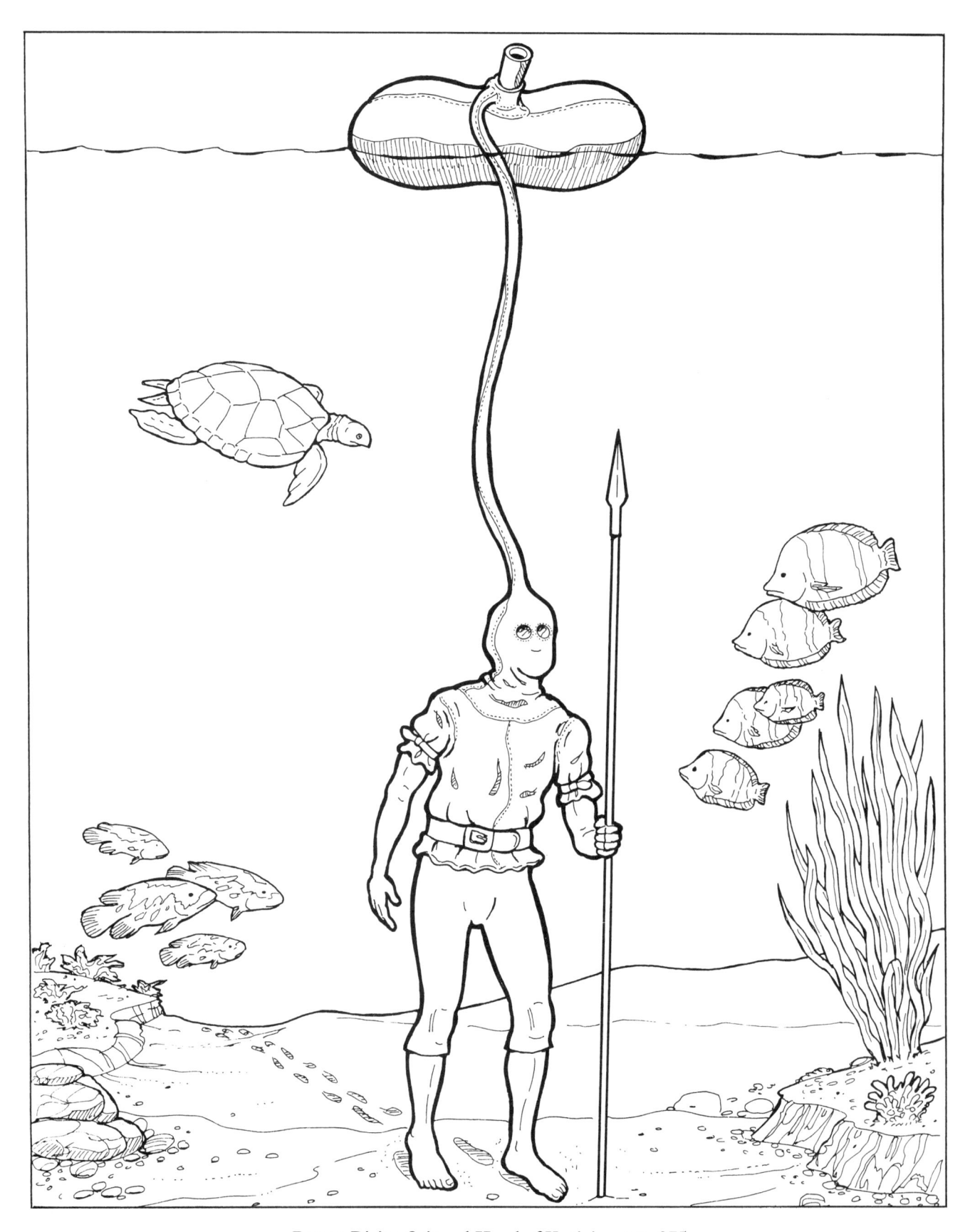

Roman Diving Suit and Hood of Vestigius, A.D. 375

Another early effort to devise an underwater diving system was made by the Roman inventor, Vestigius. His idea was to have a closed system, sealing out the water from the diver's head. The oiled-leather hood worn by the diver was attached to an inflated bag floating on the surface of the water. The air would travel from the surface down to the diver by a flexible tube. Unfortunately, this system did not allow for the efficient expulsion of the carbon dioxide gas exhaled by the diver, which could result in possible suffocation. Because of this critical shortcoming as well as some others, Vestigius' system was not developed any further.

Borelli's Diving Suit, 1660

Various unsuccessful attempts at underwater diving gear were made during the Middle Ages—a period roughly spanning A.D. 400 to 1475. The centuries preceding the eleventh are often known as the "Dark Ages," a reference to the rejection of reason and scientific inquiry in favor of superstition and ignorance. No significant advances in underwater exploration were made during this period of decline. Technology and science were revived beginning around 1475, during the period commonly called the Renaissance, which literally means rebirth.

Shown above is an example of a diving suit devised by an Italian inventor named Borelli. Developed around 1660, the apparatus was a closed, self-contained system. It consisted of a large air-filled, oiled-leather bag covering the diver's head. A glass viewing port was installed to provide forward vision. The oiled-leather suit included mittens and boots with a crude attempt at swimming fins. A tube to aid in "exhausting" the carbon dioxide gas expelled by the diver was attached to the large air bag. However, since there was no method of regulating the gas, this did not function very efficiently.

The diver also carried a metal tank and a weight to provide a means of overcoming the buoyancy of the air-filled bag. This metal tube—called "ballast" in underwater terminology—could be filled or emptied with water to change its weight. This would allow the diver to rise or sink in the water at will.

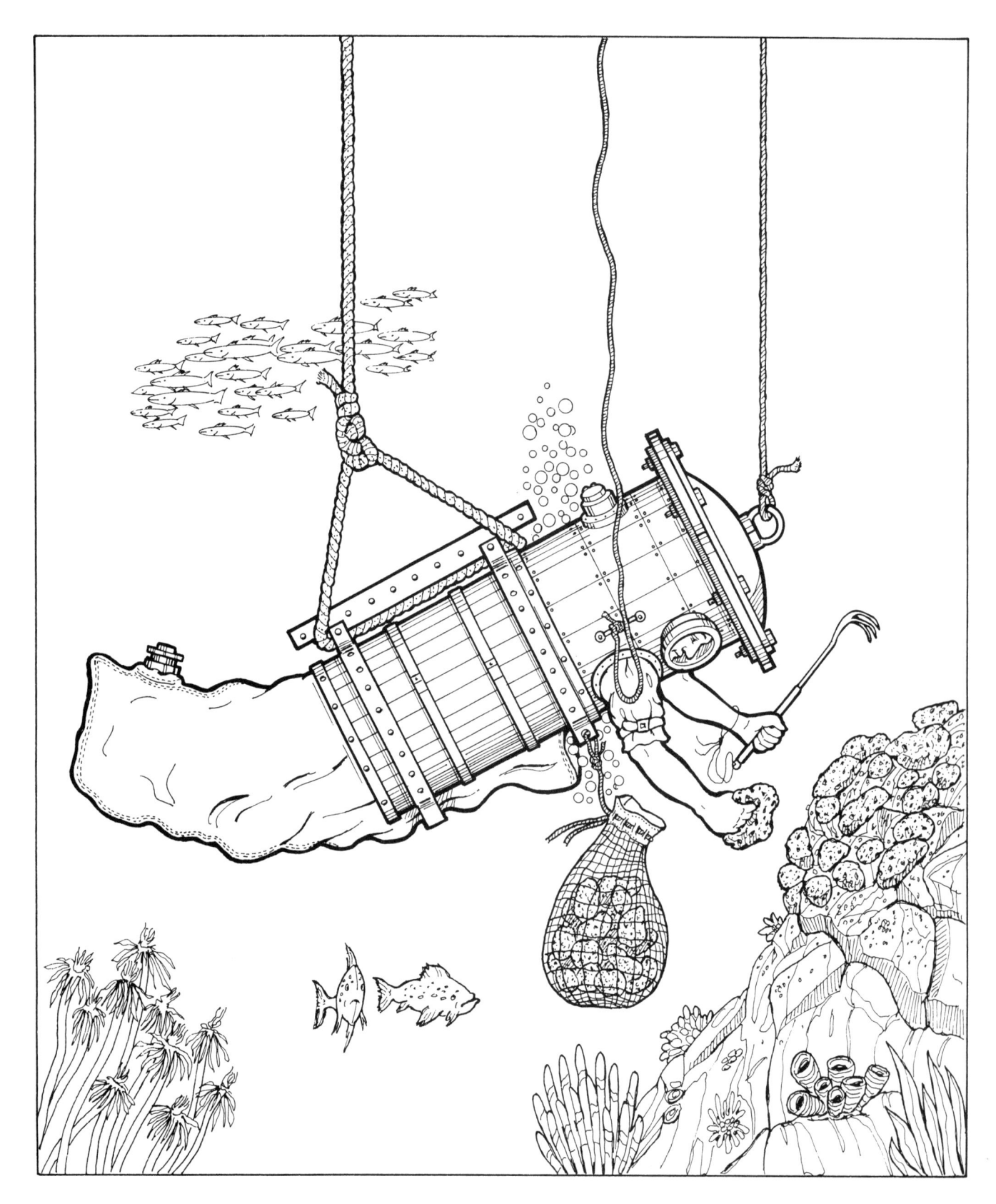

Diving Machine of John Lethbridge, 1715

Another interesting attempt at diving apparatus was invented by Englishman John Lethbridge in 1715. The upper portion of the system consisted of a metal and wood barrel, while the lower end was made of oiled leather. Within the barrel, a glass viewing port and armholes with sealed cuffs were provided. The entire mechanism, known as a "diving machine," was raised and lowered by ropes operated from a surface ship. An emergency rope near the diver's arm could be tugged to signal trouble. The lower portion of the apparatus included a vent to release carbon dioxide, but without a method of controlling this process, it wasn't very effective. Nonetheless, Lethbridge's invention was said to have been used successfully for many years, salvaging several valuable cargoes from shipwrecks. The diver depicted above is collecting sponges, a primitive sea animal.

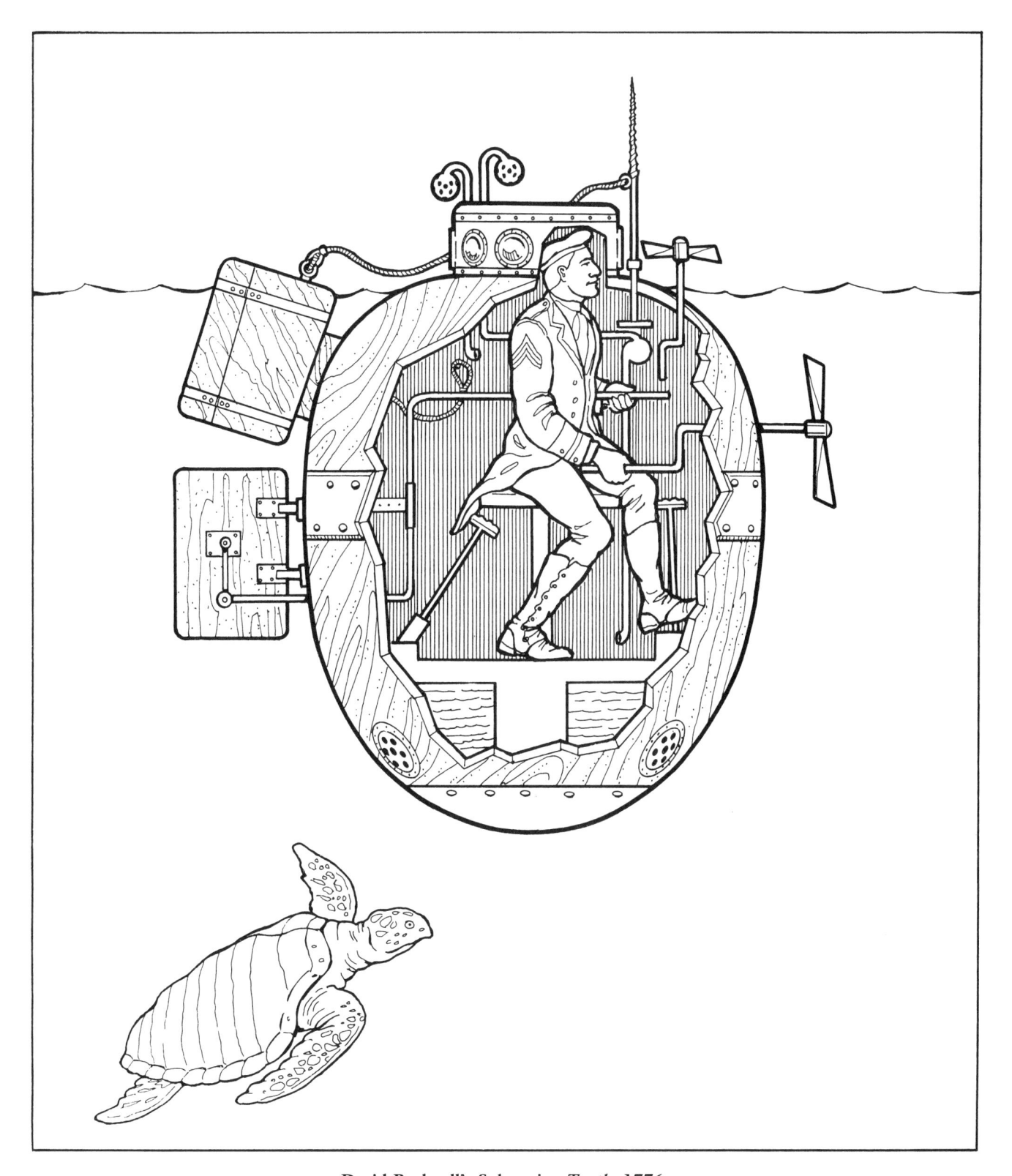

David Bushnell's Submarine *Turtle*, 1776

The first use of a submersible vessel in warfare occurred during the American Revolutionary War. Designed by American inventor David Bushnell in 1776, the egg-shaped, one-man craft was constructed of wood, iron, and copper. Its features included floodable ballast tanks, hand-cranked propellers to provide forward and downward motion, a steering rudder, air intake snorkels, and portholes. The *Turtle* carried a detachable gunpowder bomb designed to be affixed to the hull of an enemy ship by means of a screw mechanism.

On September 6, 1776, the *Turtle* launched an attack on the British warship, HMS *Eagle*, part of a force that blockaded New York harbor off the island now occupied by the Statue of Liberty. It was piloted by colonial army Sergeant Ezra Lee, who successfully maneuvered the ungainly vessel against the hull of the British warship without detection. Lee was unable to attach the bomb, however, due to the copper plating on the *Eagle*'s hull. The bomb floated away from the British warship, eventually exploding harmlessly. Sergeant Lee made a safe retreat even though the attack failed. For its time, the *Turtle* was quite an advanced undersea vessel.

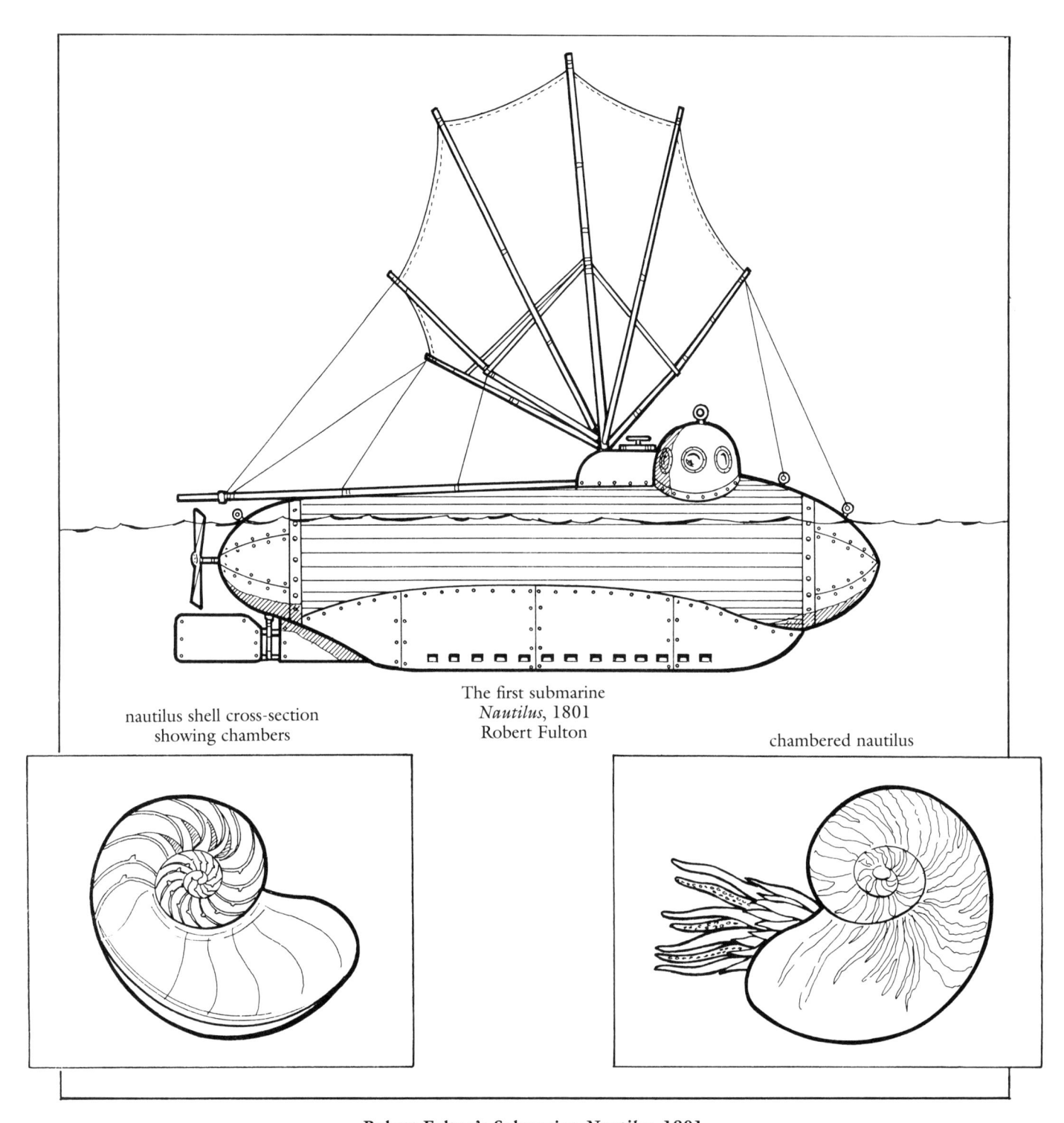

nautilus shell cross-section
showing chambers

The first submarine
Nautilus, 1801
Robert Fulton

chambered nautilus

Robert Fulton's Submarine *Nautilus*, 1801

The name "Nautilus" has been given to a number of well-known and historically significant undersea vessels. It derives from a species of sea animal—distantly related to the octopus and squid—with a spiral, chambered shell. The nautilus is the only member of the cephalopoda class that has a hard shell. The first submarine to be christened *Nautilus* was created by American engineer Robert Fulton, but many submarines both real and fictional would be designated *Nautilus* over the next 150 years, including the world's first atomic-powered submarine.

In 1797, Robert Fulton was contracted by the French Emperor Napoleon Bonaparte to design and construct an underwater boat. The result—Fulton's *Nautilus*—included many features found in modern submarines. Fabricated from wood and copper-plated iron, it had floodable ballast tanks, diving planes to control descent and ascent, and a "conning" (control) tower. The vessel was powered by a hand-cranked propeller for submerged travel, and had a collapsible sail for surface propulsion. Holding sufficient air to keep four men alive and two candles burning for three hours underwater, the submarine later contained a tank of compressed air.

Fulton successfully demonstrated the capabilities of his submarine for the French government in 1801. Although the boat stayed submerged for over an hour at a depth of twenty-five feet, it never went into operational service and the project was eventually abandoned. Robert Fulton would later go on to build one of the world's first commercially successful steam-powered surface ships, the *Clermont*, in 1807.

Augustus Siebe "Open System" Diving Apparatus, 1819

In 1819, the first reasonably practical diving gear was invented by Augustus Siebe, a mechanic from Germany who settled in England. It consisted of a copper and brass helmet with a glass face-plate. Air was pumped from the surface vessel to the diver through a flexible hose. The helmet was attached to a rubberized canvas jacket that had a bell-shaped open bottom so that the dangerous carbon dioxide gas expelled by the diver would vent from this opening (hence, "open system"). Weights were also fastened to the jacket to neutralize the diver's buoyancy and stabilize him into an upright position on the sea bottom, which was the necessary position for the proper functioning of the air supply device.

Compressed Air Tank Diving Suit of W. H. James, 1825

Another early diving suit, and one which represented a significant step toward the development of SCUBA (the acronym for Self-Contained Underwater Breathing Apparatus) was invented by the Englishman, W. H. James, around 1825. This system differed from the diving apparatus of Augustus Siebe in that the air reservoir was contained in a metal tank worn by the diver. The tank fed air to a copper helmet with a glass viewing port. Due to an inefficient method of exhausting carbon dioxide, this type of underwater gear did not enjoy widespread use.

Augustus Siebe "Closed System" Diving Suit, 1838

Pioneering undersea inventor Augustus Siebe—who came to be known as the "father of helmet diving"—perfected his "closed" diving system in 1838. It has been used as a model for all "hard hat" diving apparatus since that time, with many diving suits adapted from his original concept still in use.

The closed system diving suit was watertight and consisted of a copper and brass helmet with glass viewing ports. Like his previous diving gear, air was pumped from a surface ship to the helmet, which was fastened to a metal collar worn on the shoulders. This was, in turn, part of a rubber and canvas suit covering the entire body, and sealed at the cuffs and ankles. Lead weights hung from the metal collar, and lead-bottomed boots helped stabilize the diver in the water.

The secret to the success of this system was Siebe's invention of a valve in the helmet that efficiently discharged the carbon dioxide exhaled by the diver. The so-called "Siebe Improved Diving Dress" was adopted as the standard diving dress by Britain's Royal Engineers, and it became the principal method used in underwater exploration and activity until the advent of the SCUBA system in 1943.

Articulated Metal Diving Suit, 1850

Another less successful attempt at an underwater diving system was the metal suit depicted above. Developed around 1850, it was essentially a suit of underwater armor, very similar in appearance to both the armored suits worn by medieval knights and the pressure suits worn by modern astronauts. It consisted of a helmet, breastplate, abdominal plate, and arm and leg sleeves made from copper and brass. These pieces were connected by rubberized canvas sections that were folded or pleated to allow movement. A flexible hose from a surface vessel pumped air into the mid-section of the suit. A tube to aid in expelling carbon dioxide was also attached. This method was not nearly as efficient as the helmet valve of Augustus Siebe's closed system, and never came into widespread use.

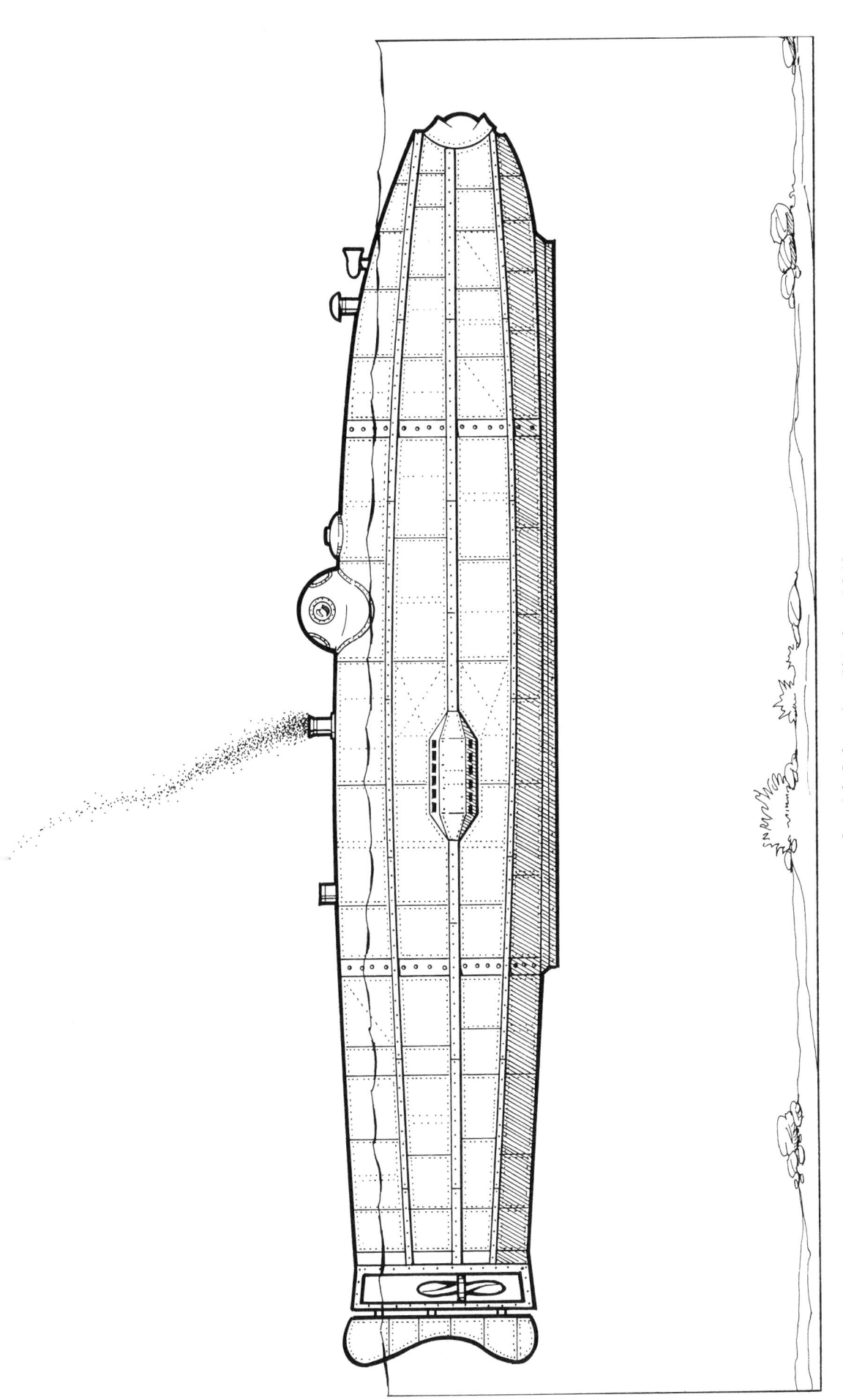

Spanish Submarine *El Ictineo*, 1862

The Spanish navy has roots as a sea power that date back to the 1500s—the "Age of Conquest" of the New World. The Spanish Armada was the most powerful naval fleet in the world during this era. Several centuries later, a number of early submarines were developed by Spain for its naval force. One of these was the *El Ictineo*, or "the submarine vessel."

The craft was designed by engineers Narcisco Monturiol and Cosmo Garcia around 1859. Built and tested from 1860 through 1862, *El Ictineo* underwent sixty successful dives during its sea trials. In many of its features, *El Ictineo* was a very modern submarine. These included a double hull for added strength against the crushing force of water pressure below the surface, ballast tanks to aid in diving and ascent, a steam powerplant turning a propeller, and a mechanical-chemical device to produce oxygen. It was armed with a single cannon, as well as numerous screw augers to bore holes into the bottoms of enemy ships. Despite *El Ictineo*'s modern design and successful operation, no other submarines of this type were constructed by the Spanish.

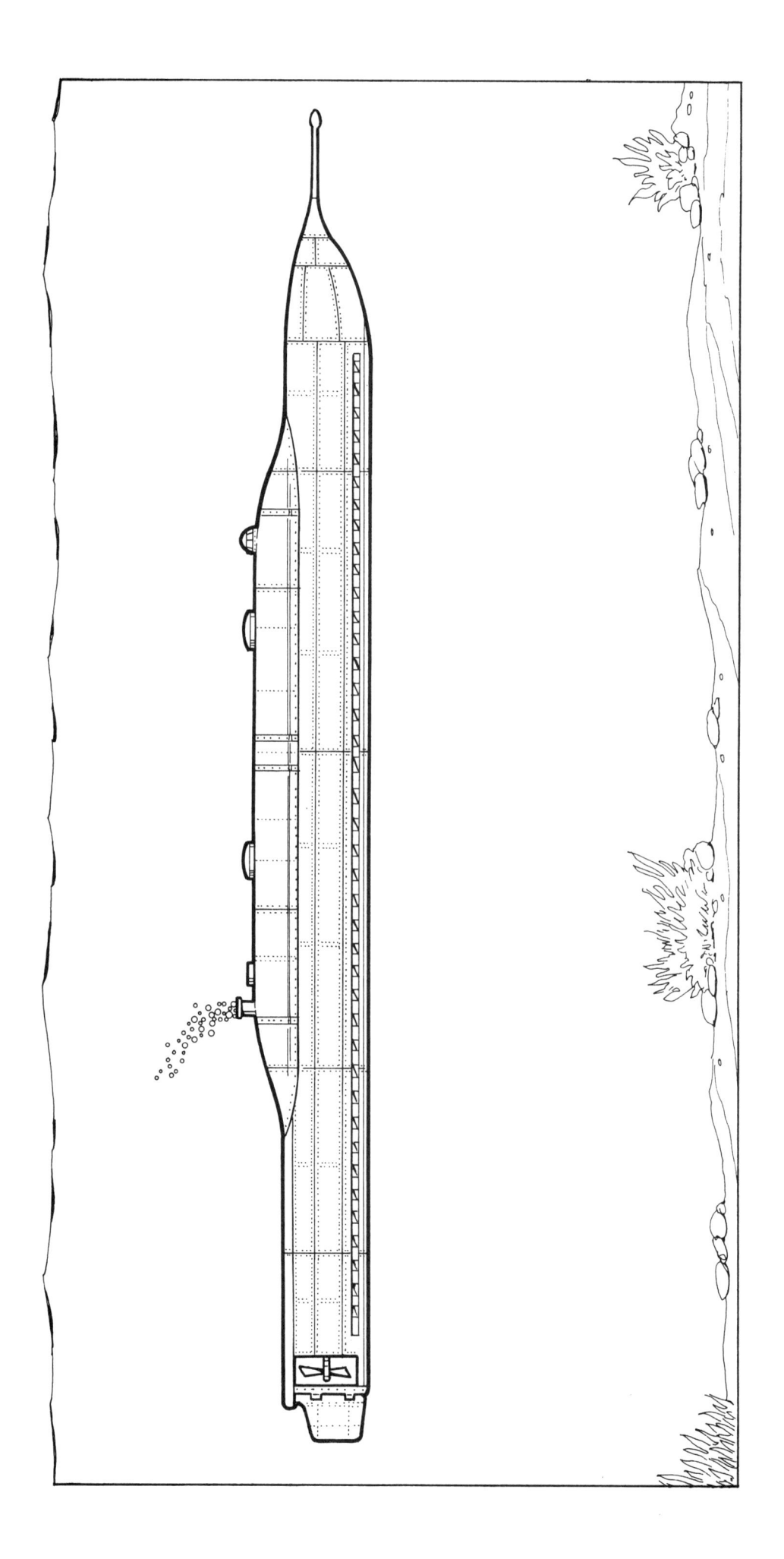

French Submarine *Le Plongeur*, 1863

Another early submarine prototype was the sleek-looking French vessel, *Le Plongeur* (the diver), launched in 1863. This submarine had an eighty-horsepower compressed air-powered engine turning a screw propeller. Compressed air was safer than steam power for underwater travel, but it made for limited range and speed, driving *Le Plongeur* at only four knots per hour submerged. (A knot is a measurement of speed equivalent to one nautical mile.) Although these early prototypes of underwater vessels were not widely used or operated, they served as platforms for the future development of more modern submarines.

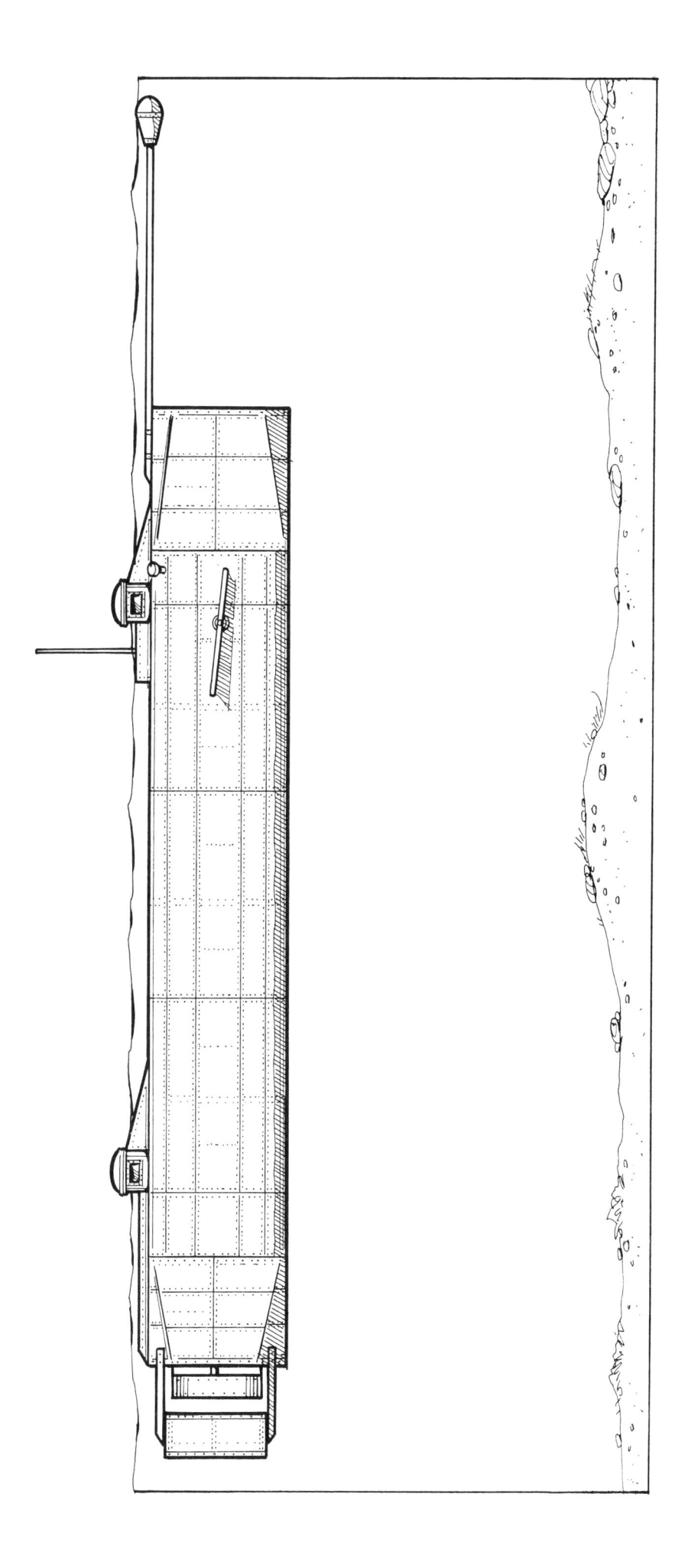

Confederate States of America Submarine, CSS *Hunley*, 1864

The American Civil War saw the use of both iron-clad surface battleships and submarines for naval warfare during that conflict. The Confederate states launched their first submarine, the CSS *Hunley*, in 1864. It was basically a cylindrical iron boiler driven by a hand-cranked propeller turned by its eight-man crew. It was equipped with ballast tanks and diving planes, and carried a single "spar-mounted" torpedo bomb filled with ninety pounds of gunpowder. (Spar is a nautical term for a long wooden pole or mast.)

In an attack on the Union steam sloop *Housatonic*, the CSS *Hunley* became the first submarine warboat to sink an enemy vessel. On the night of February 17, 1864, the *Housatonic* was anchored just outside Charleston harbor when the *Hunley* maneuvered close enough to explode its torpedo against the side of the Union ship. The explosion ignited the ammunition magazine aboard the *Housatonic*, sinking the vessel. Unfortunately for the crew of the *Hunley*, the waves from the explosion swamped their boat through an open hatch, causing the *Hunley* to sink with all hands on board.

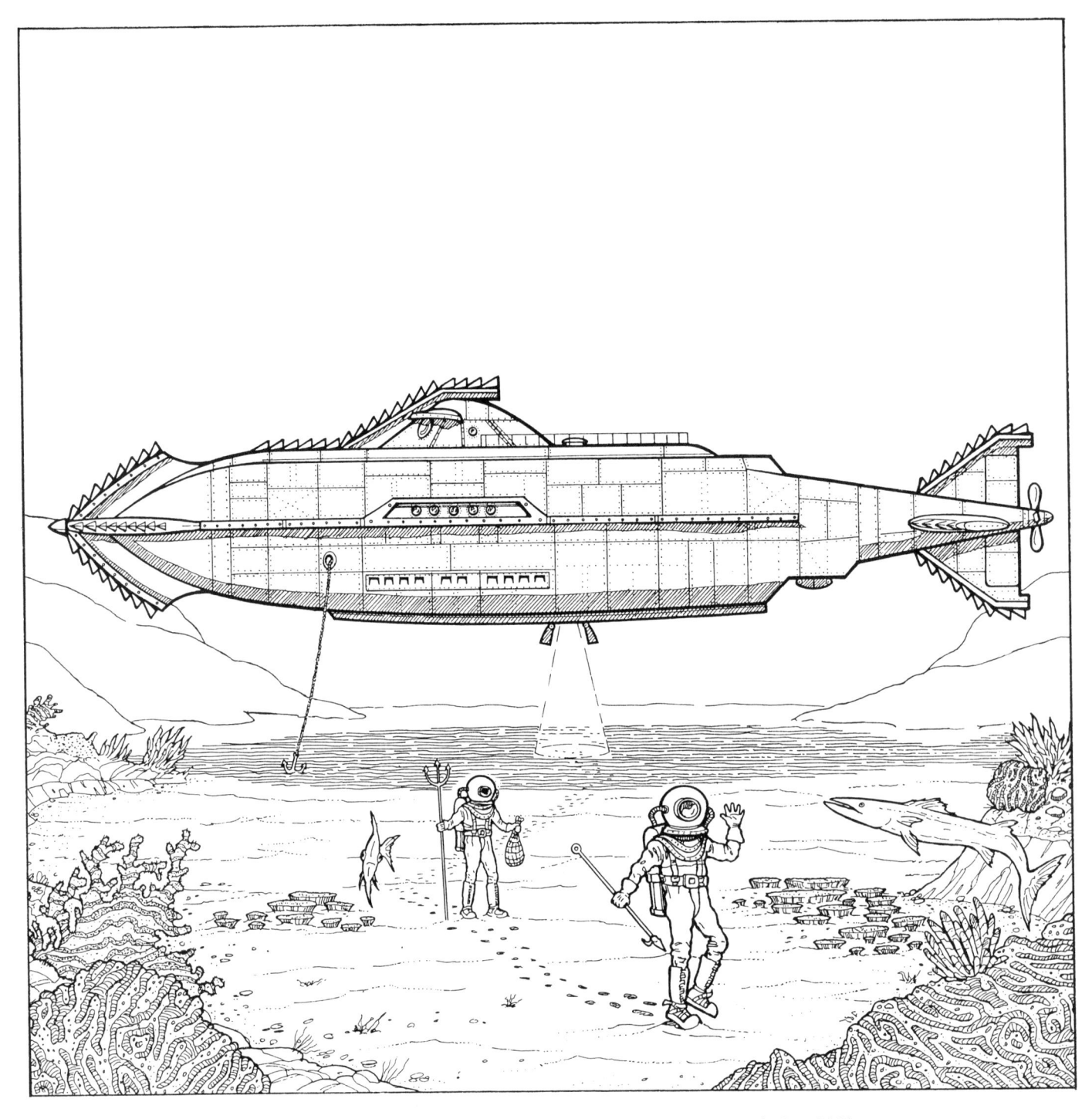

Jules Verne's *Nautilus* from *Twenty Thousand Leagues under the Sea*, 1870

No history of the development of submarines would be complete without a reference to the fictional undersea vessel, the *Nautilus*, from the 1870 Jules Verne novel *Twenty Thousand Leagues under the Sea*. Verne has often been called the "father of modern science fiction." In the books that he wrote during the last half of the nineteenth century, Verne was eerily accurate in predicting scientific and technological developments well ahead of his time. Verne's adventure stories featured detailed descriptions of futuristic machines that included gigantic airships which could fly great distances at high speed, spacecraft capable of circumnavigating the moon, and of course, the fabulous submarine *Nautilus,* able to travel deep beneath the oceans for many days powered by a mysterious and unknown form of energy.

The *Nautilus* that Verne envisioned was in many respects a modern nuclear submarine complete with ballast tanks, diving planes, and speedy underwater propulsion from an energy source described as the "veritable dynamic force of the Universe." In Verne's book, the submarine's fictional creator was the enigmatic Captain Nemo, a self-styled outcast and tormented genius who unleashed this new form of energy to power his underwater vessel. Together with his crew of ex-convicts and former slaves, Captain Nemo and the *Nautilus* prowled deep beneath the oceans of the world like some mythical sea monster, seeking and destroying warships of all nations. As created by Jules Verne, Captain Nemo and his futuristic *Nautilus* have added a colorful page to the history and lore of the submarine. And it seems only fitting that the world's first nuclear-powered submarine, the U.S. Navy SSN-571 launched in 1955, should bear the historic name *Nautilus*.

Common Dolphins, 10 feet

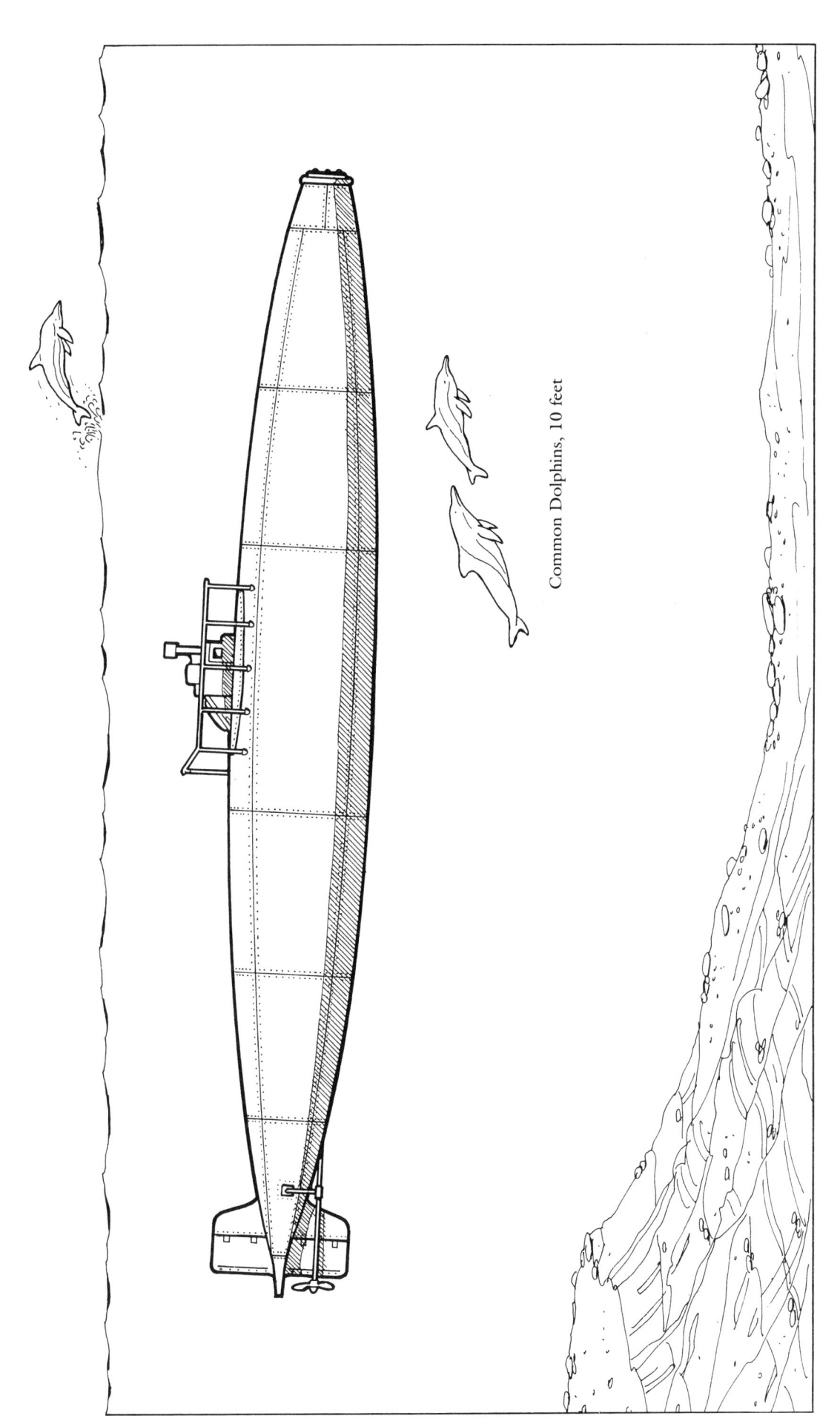

Spanish *Peral* Class Submarine, 1888

One of the most successful prototypes for late nineteenth-century submarines was the vessel depicted above that was designed by the Spanish navy Lieutenant Isaac Peral. His boat—in naval terminology, a submarine is always referred to as a "boat," rather than a "ship," regardless of the vessel's actual size—featured propulsion by battery-driven electric motors, a system later used by many modern submarines.

The diving and surfacing of the boat were controlled by both ballast tanks and vertical propellers mounted at the bow and stern.

The vessel carried three torpedoes that could be fired from a forward-mounted tube covered by the blunt cap at the bow. The weight of a submarine or any ship is called its "displacement." This refers to the amount of water that the vessel displaces as it sits in the water. Peral's submarine had a displacement of eighty-seven tons. It could travel at ten knots on the surface, and was seventy feet long. Although launched and successfully tested during 1888, only one *Peral* class submarine was built.

Pilot Whales 20 feet, 15 feet

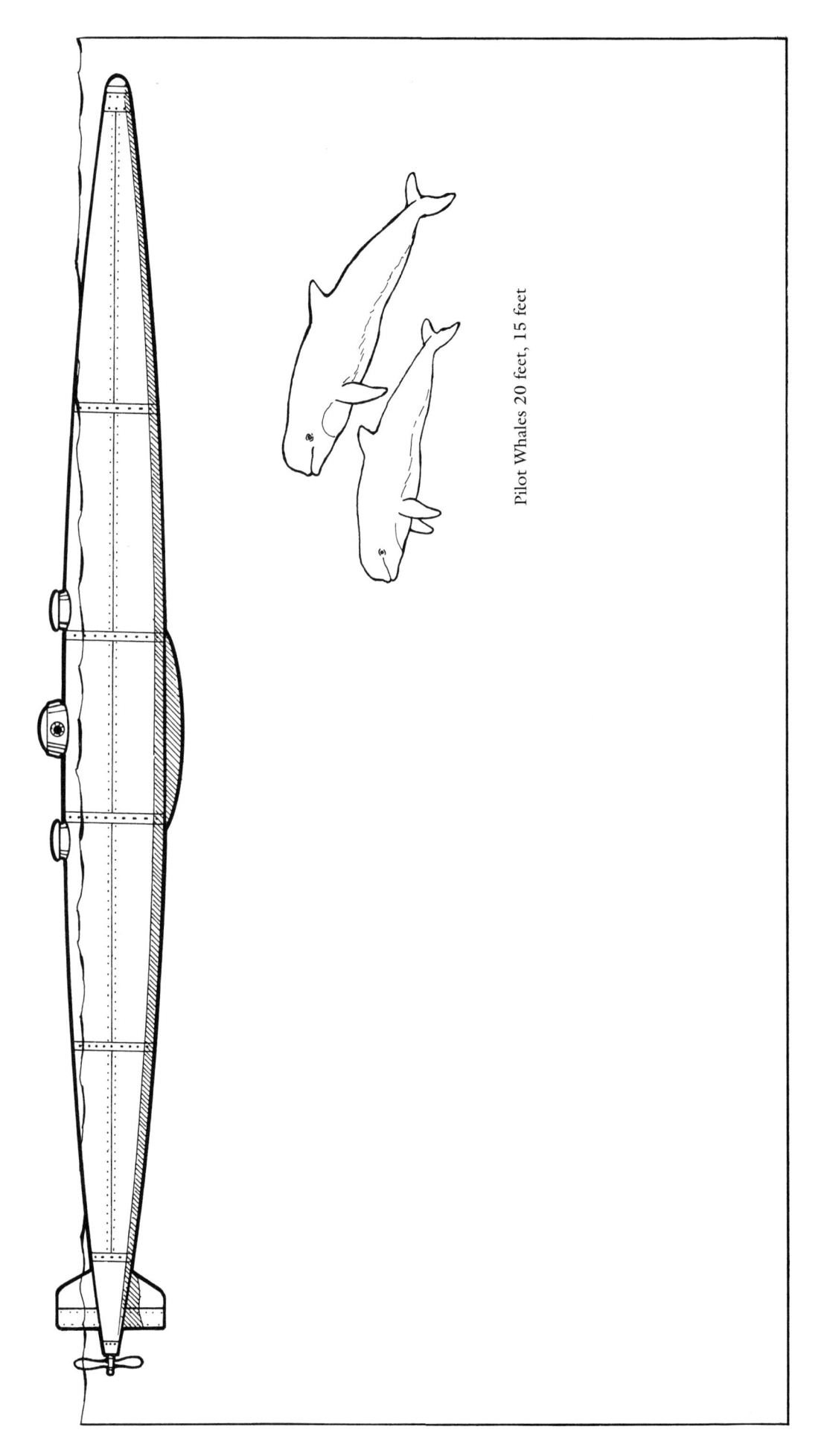

French Submarine *Gustave Zédé*, 1891

The sleek-looking French submarine *Gustave Zédé*, pictured above, was an electric-powered vessel based on a successful earlier design, the *Gymnote*. Both vessels used batteries to power diesel-electric motors that drove a single stern-mounted propeller. The *Gustave Zédé* had one torpedo tube at the bow and carried three torpedoes. The boat made over 2,000 dives during its operational service. It was finally decommissioned in 1909.

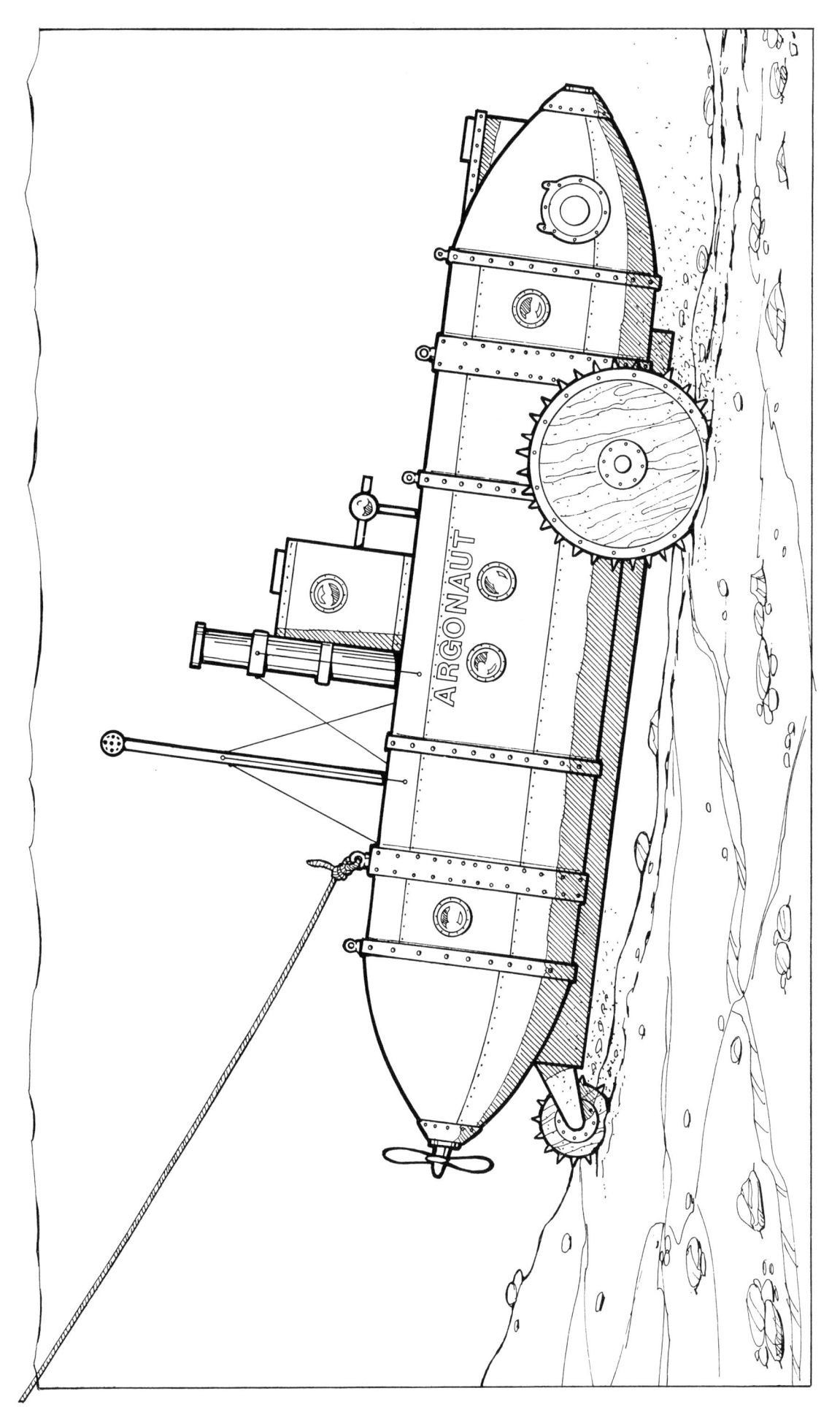

American Submarine *Argonaut*, 1897

An American engineer and naval architect, Simon Lake—who came to be known as the "father of the modern submarine"—designed a submersible bottom-crawler boat in 1897. It was named the *Argonaut*, after the crew of the legendary ship *Argo* from the Greek myth of Jason and the Argonauts.

The *Argonaut* was intended as an undersea research craft, designed to send out divers for exploration of the seabed. It had a thirty-horsepower engine driving the rear wheels and a forward-mounted propeller. The front wheel was used for steering. An air chamber was included which allowed divers to leave and return to the boat. It carried a crew of five and once journeyed from Norfolk, Virginia to New York City in 1898 through heavy storms, making it the first submarine to operate successfully in the open sea.

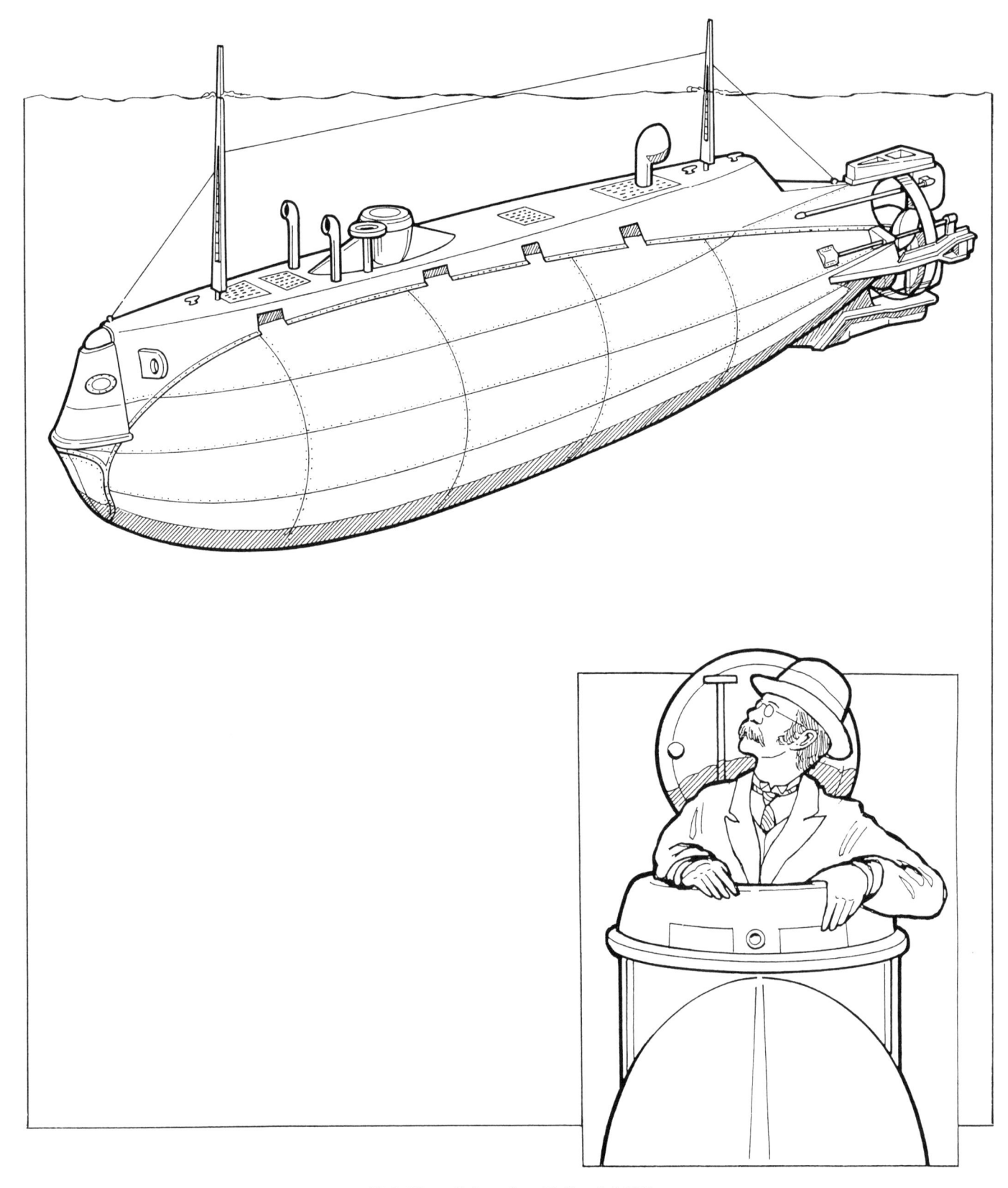

U.S. Navy Submarine *Holland*, 1897

A milestone in the development of the modern submarine was the USS *Holland*, launched in 1897. Named after the boat's designer, Irish-American engineer John Holland, this undersea vessel was the first submarine to go into widespread operational use within the U.S. Navy, starting in 1900. The *Holland* was equipped with a gasoline engine for surface travel, and electric motors for underwater propulsion. The maximum speed of the fifty-three foot long, sixty-four ton craft was eight knots surfaced, and five knots submerged. The *Holland*'s operational depth limit was seventy-five feet. The smooth teardrop-shaped hull was advanced for its time, and the boat relied on ballast tanks and stern diving planes for underwater descent and ascent.

The basic design of the *Holland* was used as a model for construction of the early submarines of both Great Britain and Japan, and it also served as a training vessel for the United States Navy until it was finally scrapped in 1913.

Humpback Whale, 50 feet

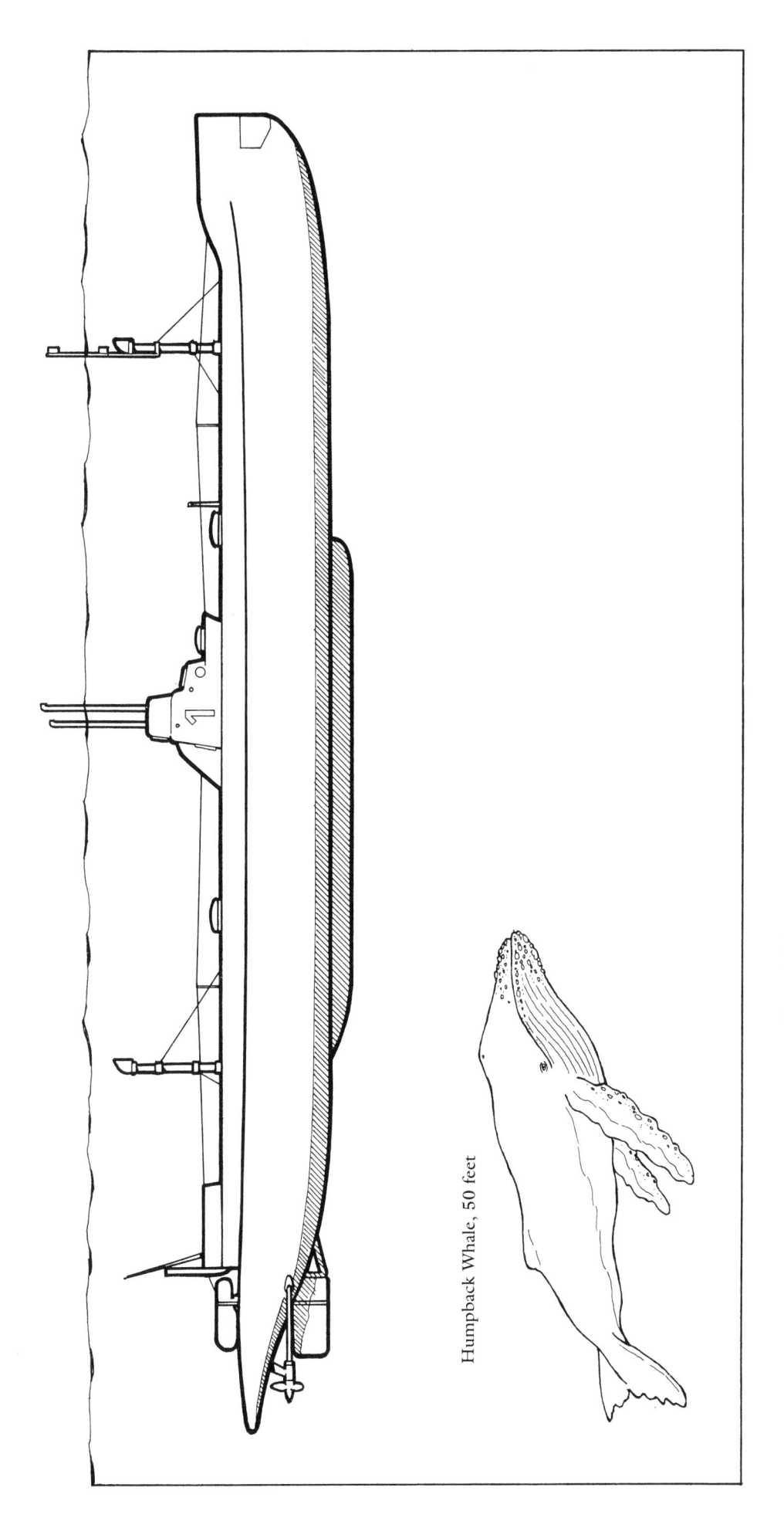

Imperial German Navy U-1 (*Unterseeboot*) Submarine, 1906

The first German submarine to go into operational service was the 238-ton U-1, launched in 1906. It became the model for a large fleet of submarines constructed by Germany during the period leading to the World War I era (1914–1918). The U-1 was equipped with two kerosene engines for surface travel, and battery-driven electric motors for undersea propulsion. These engines drove two stern propellers—called "screws" in nautical terminology—which gave the boat a top speed of eleven knots on the surface and nine knots submerged.

The U-1 carried three torpedoes fired through her single bow-mounted torpedo tube. The boat was 138-feet long, had a "beam" (width) of thirteen feet, and a maximum diving depth of 100 feet. The vessel was equipped with a telescoping "periscope" to guide her during underwater attack runs. Although the U-1 did not see combat during World War I, it was instrumental in training submarine crews for that conflict.

19

1900   USS Holland, 53-feet long, 64 tons

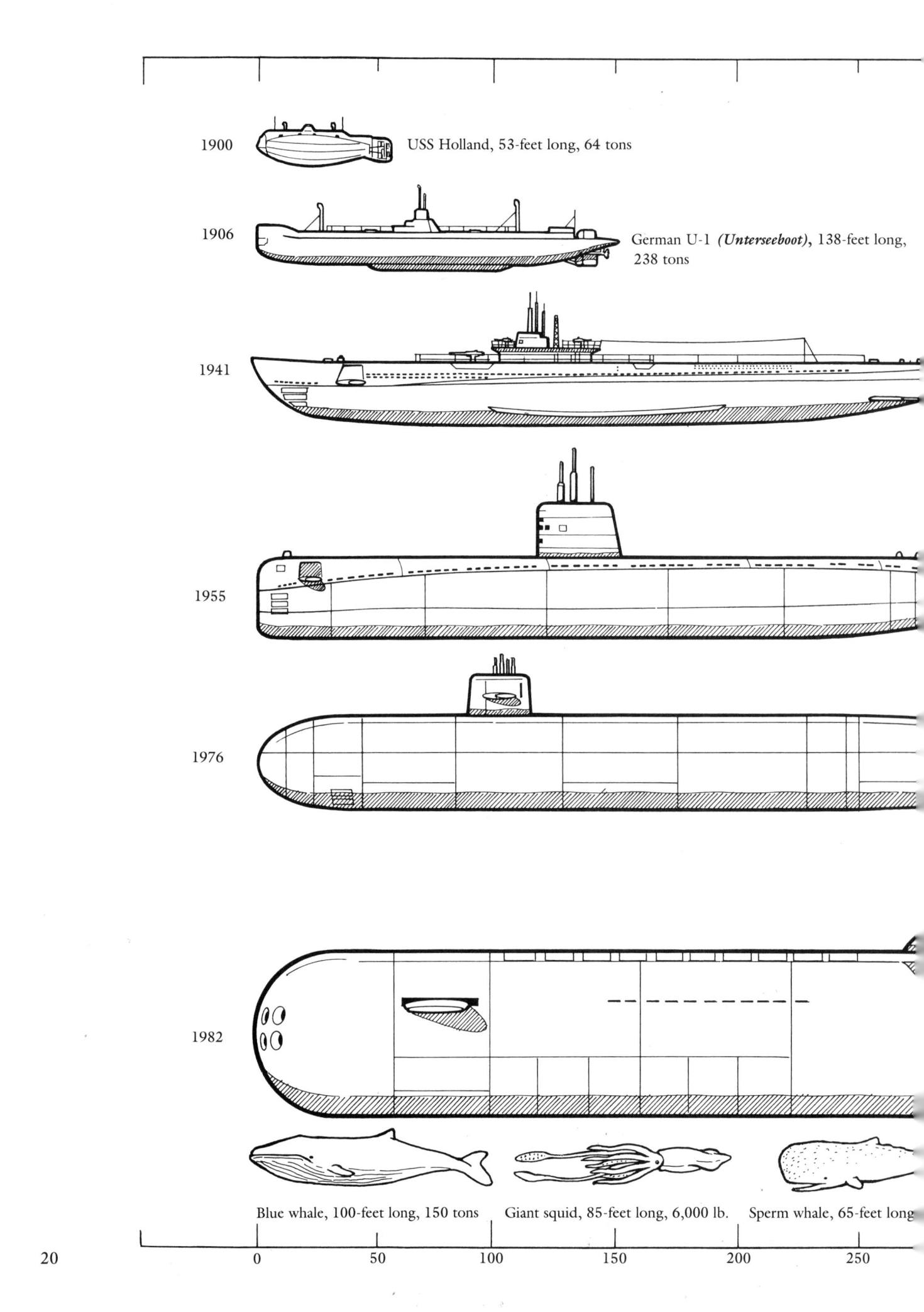

1906   German U-1 (*Unterseeboot*), 138-feet long, 238 tons

1941

1955

1976

1982

Blue whale, 100-feet long, 150 tons   Giant squid, 85-feet long, 6,000 lb.   Sperm whale, 65-feet long

0   50   100   150   200   250

American World War II *Gato* class submarine, 311-feet long, 1,825 tons

American atomic submarine, SSN-571 *Nautilus*, 323-feet long, 3,674 tons

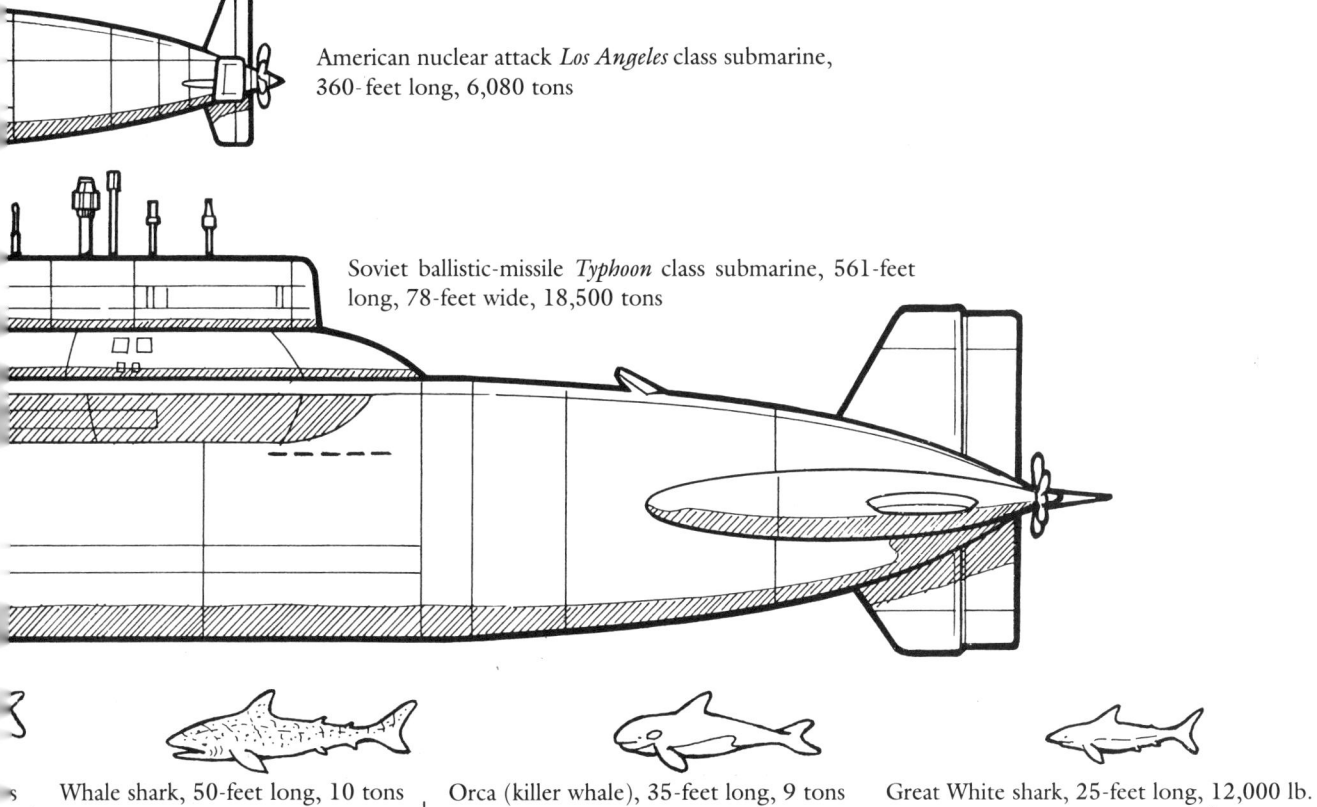

American nuclear attack *Los Angeles* class submarine, 360-feet long, 6,080 tons

Soviet ballistic-missile *Typhoon* class submarine, 561-feet long, 78-feet wide, 18,500 tons

Whale shark, 50-feet long, 10 tons   Orca (killer whale), 35-feet long, 9 tons   Great White shark, 25-feet long, 12,000 lb.

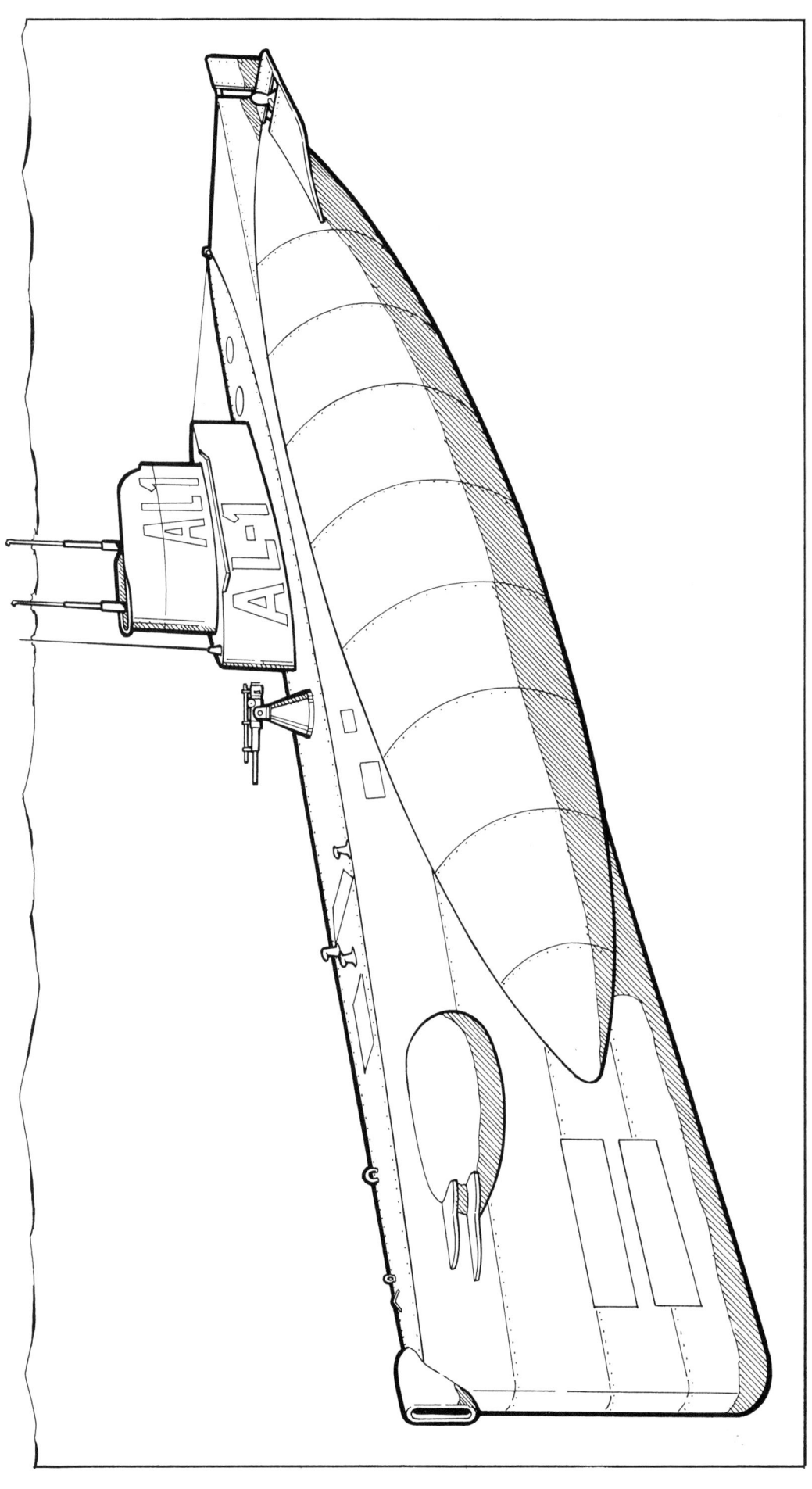

American World War I L-1 Fleet Submarine, 1917

As the United States entered World War I in 1917, a new class of submarines were becoming operational within the U.S. Navy. These were the "L" class boats, as depicted above. The vessels in this class were 167-feet long, had a beam of eighteen feet, and displaced 450 tons. Their maximum diving capability was 200 feet. They were powered by 900-horsepower diesel engines and 700-horsepower electric motors for surface and underwater propulsion. Top speed was fourteen knots on the surface and eleven knots submerged. The boats had an effective combat range of 3,300 miles running on the surface and 150 miles while submerged.

Their armament consisted of four torpedo tubes in the bow with a normal complement of eight torpedoes. The "L" class boats were also fitted with a deck-mounted cannon that fired a three-inch explosive shell used for engaging the enemy on the surface. Although the American "L" class boats were active in combat patrols during World War I, they were not involved in sinking any German ships. This class of submarines was decommissioned by 1923. More modern and deadly undersea warships were being designed that would make submarine combat one of the crucial elements of the next global conflict—World War II.

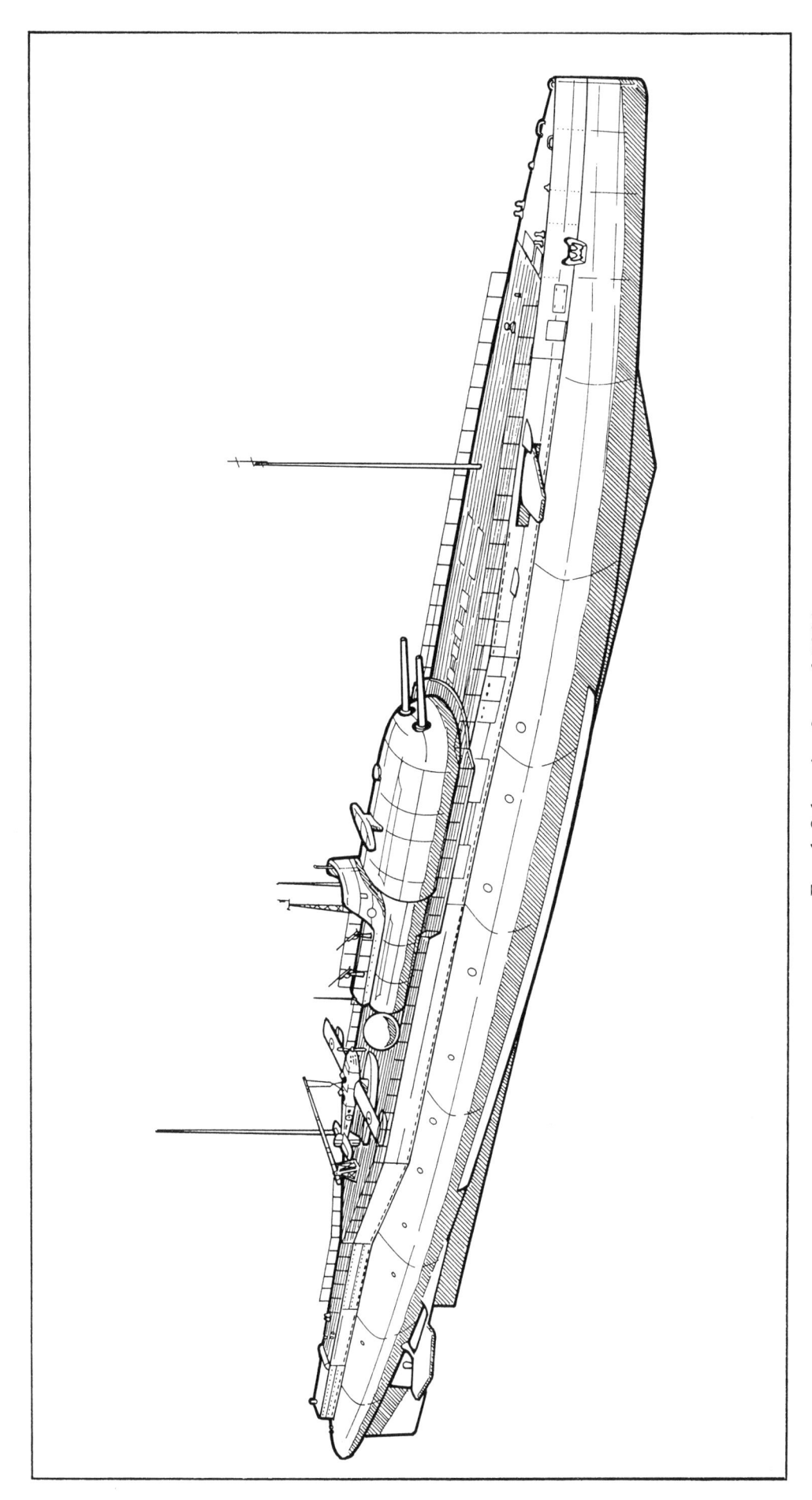

French Submarine *Surcouf*, 1934

Just prior to the beginning of World War II, the French navy constructed the *Surcouf*, a massive submarine which was the largest in the world at the time. The boat was an immense 360-feet long, had a beam of thirty feet, and displaced 3,252 tons. It had an incredible combat range of 10,000 nautical miles, and was designed for long-range operations as both an underwater attack vessel and a heavily-armed surface raider. Toward this end, the *Surcouf* mounted the largest naval guns ever installed on a submarine. Its main battery consisted of two turret-enclosed eight-inch guns which could fire an explosive shell over fourteen miles. The boat also mounted secondary deck armament of two thirty-seven millimeter cannons and two thirteen-millimeter machine guns.

The *Surcouf* was also equipped with twelve torpedo tubes for underwater attack. Four bow-mounted and four stern tubes were complemented by a rotating external mount with four tubes. A float-plane used for reconnaissance was also carried in a watertight deck hanger. A crane lifted the aircraft onto the water for launch and recovery. An unusual mixture of submarine and surface warship, the *Surcouf* never functioned very successfully in either role. While sailing to a patrol station in the Pacific in 1942, the vessel was accidentally rammed and sunk with all hands lost.

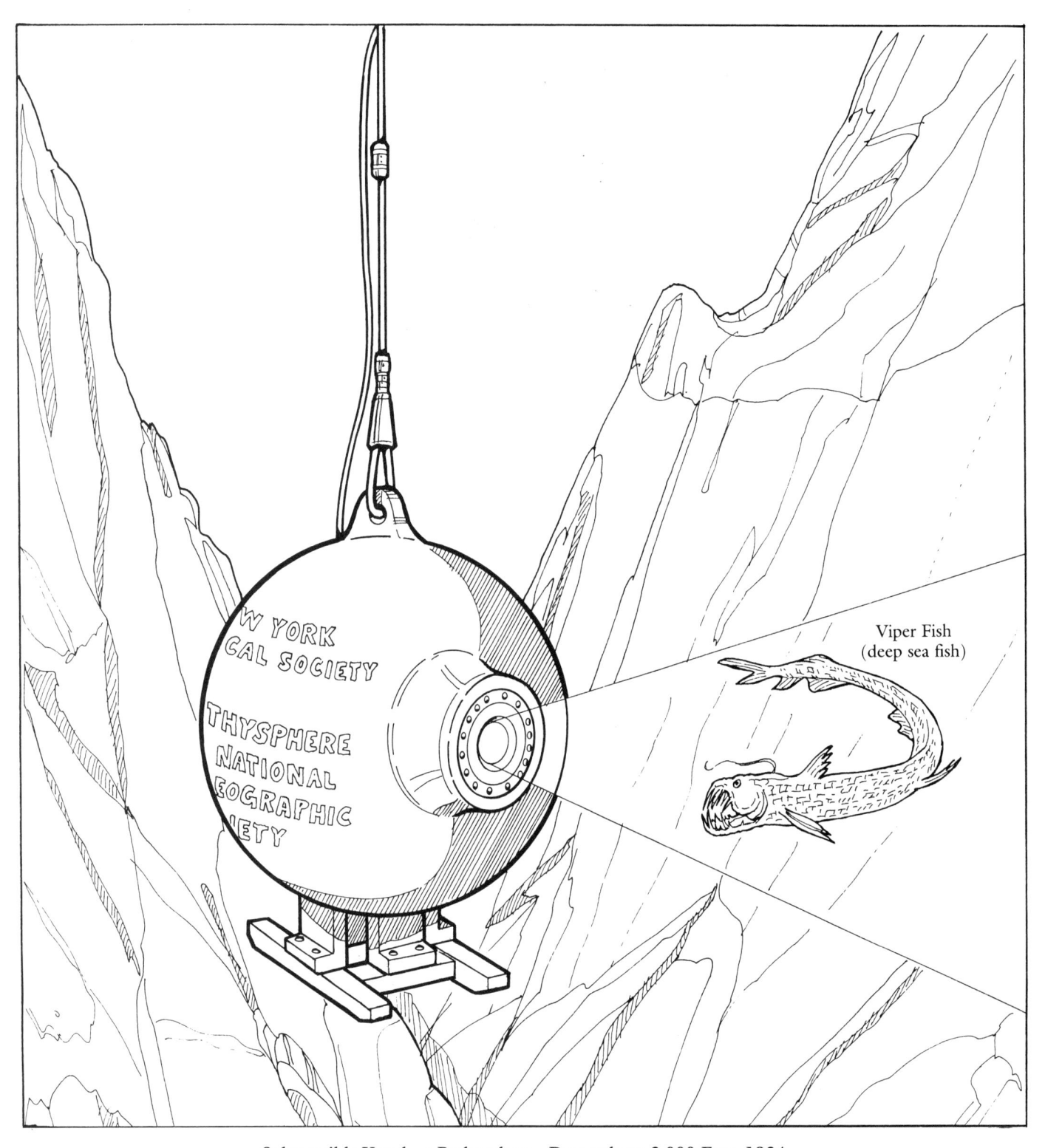

Submersible Vessel—a Bathysphere—Descends to 3,000 Feet, 1934

Underwater exploration took a major step forward in 1934 with the advent of a submersible vessel, or bathysphere. The fundamental difference between a "submarine" and a "submersible" is that a submarine is designed for prolonged periods of underwater travel, while a submersible is a vessel that can submerge for short periods over short distances.

This diving craft was essentially a heavy steel sphere built to withstand the immense pressure of deep water. It carried two passengers, scientists William Beebe and Otis Barton, to a record depth of 3,000 feet—an achievement that remained unsurpassed for fourteen years. Beebe later wrote of the experience that it was "almost superhuman, cosmic . . . [with] the long cobweb of cable leading down through the spectrum to our lonely sphere, where, sealed tight, two conscious human beings sat and peered into the abysmal darkness . . . isolated as a lost planet in outermost space."

The sphere featured a steel-doored hatch for entry and exit, and three small portholes for external viewing. The bathysphere was lowered and raised by a steel cable from a surface ship. This small and simple vessel paved the way for the more modern submersible research craft that were eventually developed in the 1960s.

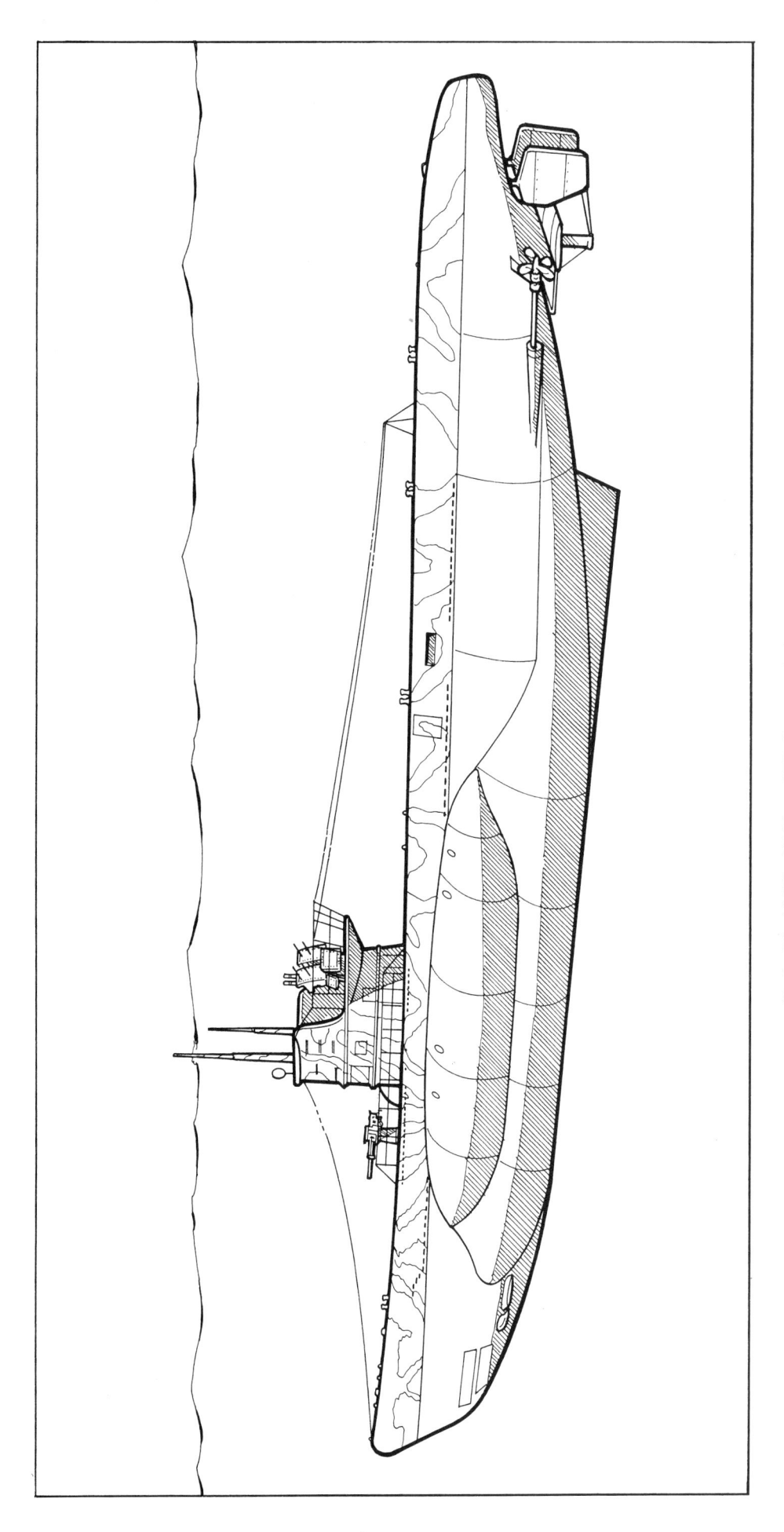

Early World War II Type VII German Submarine, 1939

With the advent of World War II in 1939, submarine technology accelerated rapidly. Wartime needs caused tremendous expenditures, research, and development in all areas of science and industry. The German "Kriegsmarine" (war navy) developed a large fleet of fast and capable ocean-going submarines for their war of aggression.

The workhorses of this fleet during the first few years of the war were the Type VII U-boats. First launched in 1936, numerous models of the Type VII saw extensive combat during the war years. They became a serious threat to the warships and merchant vessels of the Allied powers crossing the Atlantic. As weapons, troops, and other war materiel critical to the war effort were transported by sea from the United States to Great Britain, German submarines prowled undersea in multiple-boat "Wolf-packs," taking a huge toll of hundreds of sunken Allied ships.

The Type VII submarines were 221-feet long, twenty-one feet wide, and displaced 770 tons. They were powered by diesel engines and electric motors driving twin stern screws. Top speed was seventeen knots on the surface and eight knots submerged. They could reach a depth of 350 feet. The boat's deck armament consisted of an eighty-eight millimeter gun and four twenty-millimeter anti-aircraft guns. The Type VII had five torpedo tubes, four in the bow and one stern-mounted. They normally carried between eleven and fifteen torpedoes on a combat patrol. Over 700 Type VII submarines were constructed during the war years.

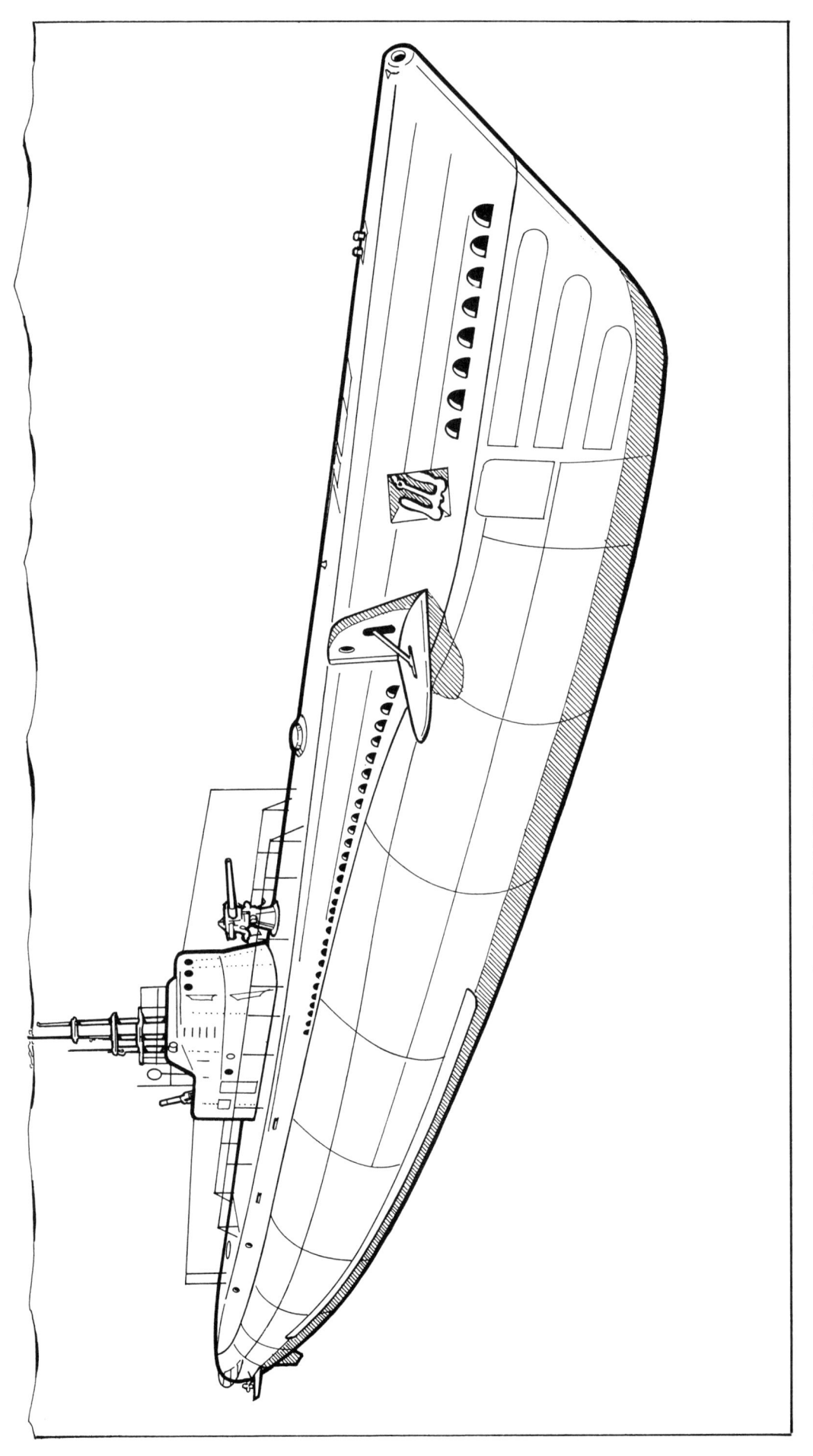

American World War II *Gato* Class Fleet Submarine, 1941

As a combatant in World War II, the U.S. Navy deployed a large fleet of efficient and reliable fighting submarines of the *Gato* class. First entering service in 1941, over 300 *Gato* class boats and its similar sisters of the *Balao* and *Tench* class, were built during the war years. These vessels formed the heart of the massive U.S. Navy undersea warfare strategy in the Pacific. Several boats of this fleet were credited with sinking over 90,000 tons of enemy shipping, contributing greatly to the successful outcome of the naval campaign.

The *Gato* class submarines were 311-feet long, had a twenty-eight foot beam, and displaced 1,825 tons. They were powered by diesel engines and electric motors that propelled them at a surface speed of twenty knots and an underwater speed of nine knots. They could submerge to a depth of 300 feet. Their deck armament by the end of the war was one five-inch gun, two forty-millimeter anti-aircraft guns, and two twenty-millimeter cannons. For underwater attack runs, they were fitted with ten torpedo tubes, six forward and four at the stern. A complement of twenty-four torpedoes was normally carried for combat patrol.

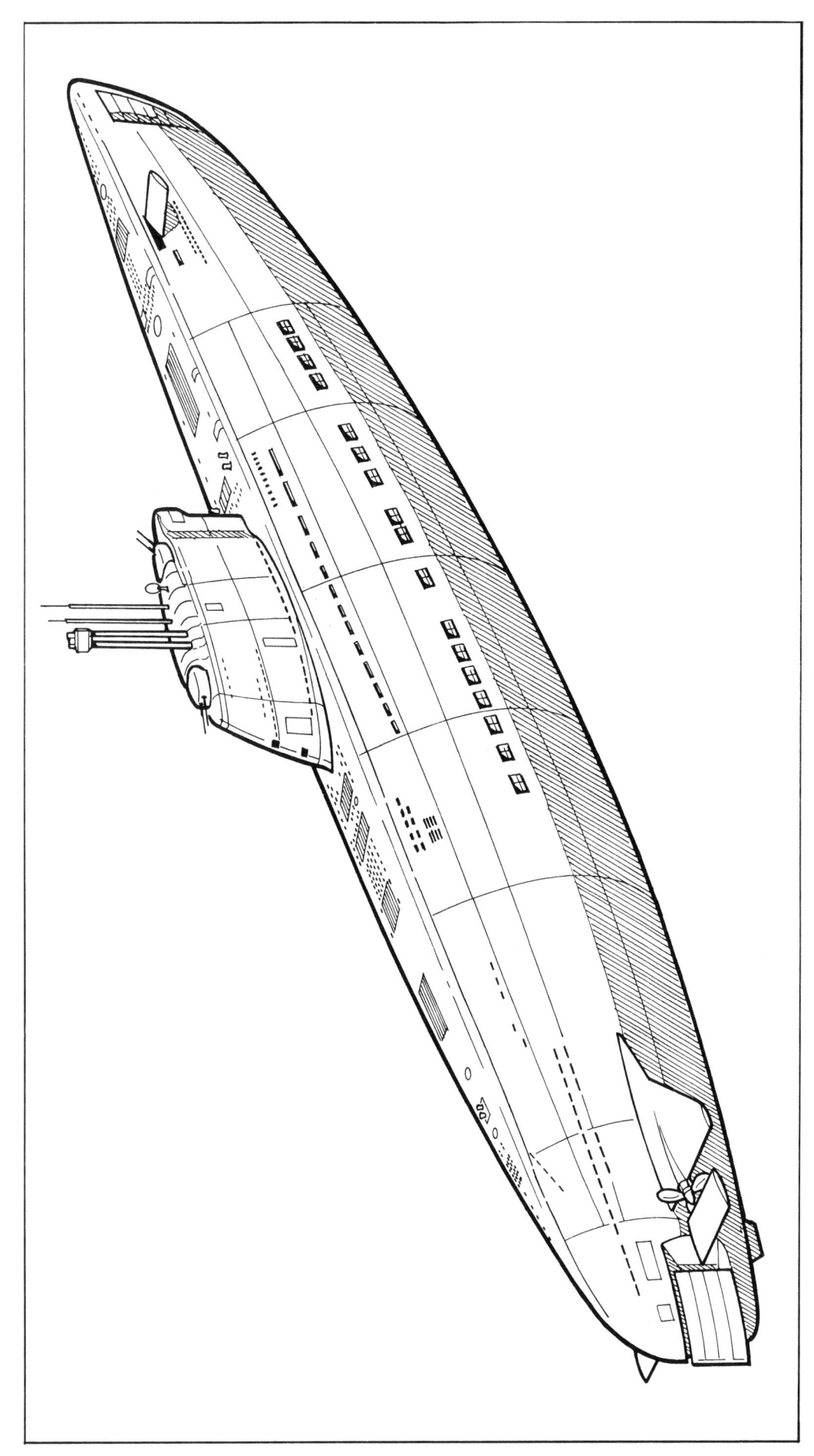

Late World War II German Type XXI Submarine, 1943

The German navy developed a very sophisticated type of submarine towards the end of World War II. This was the fast and quiet "Type XXI," first deployed in 1943. Its powerful diesel engines and newly designed electric batteries gave it exceptional speed, especially when submerged. The Type XXI was the first submarine to have a faster underwater speed—eighteen knots—than its surface speed of fifteen knots. The boats were also equipped with an advanced "snorkel" tube for replenishing the air supply while running submerged.

The Type XXI boats were fitted with six bow-mounted torpedo tubes that were able to rapidly reload from their reserve of twenty-three torpedoes. For surface action, they were equipped with four twenty-millimeter cannons mounted in two streamlined turrets. The boats were 252-feet long, displaced 1,621 tons, and could descend to a depth of 440 feet. Fortunately for the Allies, however, only a relatively small number of these advanced submarines were built. By the time they entered service, Germany was losing the war; its factories and ports were being bombed constantly, and production of these boats was slow and irregular. Had the Germans been able to produce these superior undersea warships in great quantity, Allied casualties would have been much higher and the war might have been prolonged. Like many of the Nazi regime's futuristic "wonder weapons," the Type XXI was a case of too little, too late.

Jacques-Yves Cousteau

SCUBA (Self-Contained Underwater Breathing Apparatus), 1943

Underwater exploration has followed along a dual path throughout its history. One pathway led to the advanced technology of the submarine warship, and as with many scientific and industrial developments, the aggressive demands of warfare drove this progress at a rapid pace. The other track led to the contrivance of submersible vessels for research and diving equipment for individual explorers. Of the latter, the most important was the invention and widespread use of SCUBA diving gear. SCUBA is an acronym for "Self-Contained Underwater Breathing Apparatus," and was the brainchild of French naval officer Jacques-Yves Cousteau and engineer Émile Gagnan, an engineer who specialized in control-valve design. With the successful demonstration of their apparatus in 1943, they opened up the world beneath the sea to extensive human exploration.

The SCUBA system incorporates a tank of compressed air worn by the diver, with rubber hoses that carry air from the tank to a mouthpiece held in the diver's mouth. The key to the efficiency and success of this system is the regulator valve. This device controls the air inhaled by the diver from the tank on his back, as well as the expulsion of deadly carbon dioxide (CO_2) that the diver exhales. With tanks containing enough air for several hours, the diver was free to roam and explore beneath the sea unencumbered by a tether hose from a surface vessel. With the aid of rubber swimming fins or "flippers" and water-tight face-masks, divers could at last swim with the mobility and agility of fish.

Jacques-Yves Cousteau would spend the next fifty-five years of his life developing more advanced SCUBA gear, new submersible vessels, and educating the public about the incredible ocean world he explored. He became world famous as a scientist, educator, writer and television host, and as a passionate spokesman for the protection of oceans and marine life. The foremost underwater explorer of the twentieth century, Captain Cousteau died in 1997 at the age of 87. His legacy will endure long into the future.

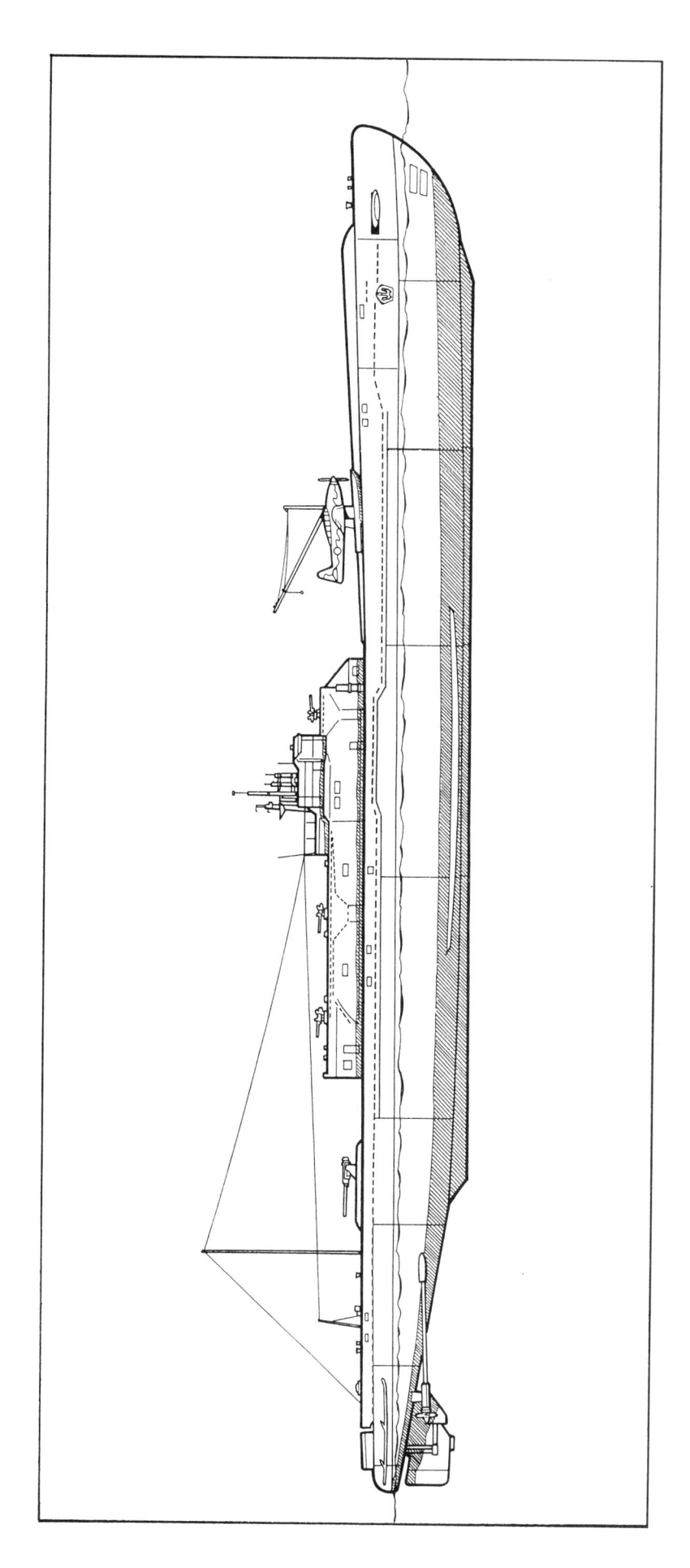

Japanese World War II "Sto I-400" Class Submarine, 1944

Toward the end of World War II, the Japanese Imperial Navy developed three aircraft-carrying submarines, the I-400 series. They were part of a desperate attempt to turn the tide of a conflict they were losing badly. The huge boats were built with the intention of sailing them within close range of the American West coast, launching their aircraft, and then bombing American cities. Like the Nazi attempts at specialized wonder weapons, these submarines failed to make a significant impact on the war. Only three boats were built, and these were difficult to operate and resupply. They were, however, the largest submarines ever constructed until the introduction of the American "Ethan Allen" class nuclear boats in the 1960s.

The I-400 was 400-feet long, had a beam of thirty-nine feet, and displaced 5,233 tons. Her watertight deck hangar could house three bomb or torpedo-carrying float planes. The massive vessel was equipped with eight torpedo tubes and carried twenty torpedoes. Her deck armament consisted of one five-inch cannon and ten twenty-millimeter anti-aircraft guns. The I-400 could reach eighteen knots on the surface and seven knots submerged. Her maximum depth capability was 330-feet. All three I-400 boats surrendered to the U.S. Navy in 1945.

circa 1950

circa 1990

"Hard Hat" Divers, 1950 and 1990

Before the invention of SCUBA gear in 1943, the metal-helmeted "hard hat" diving system was the most common version used for individual underwater exploration and activity. Utilized primarily during the first fifty years of the twentieth century, this equipment originated and developed from Augustus Siebe's original design of 1838. These divers performed underwater construction, repair and maintenance of ships, salvage operations, and marine life study.

The hard hat diver shown above on the left wears equipment typical of the kind used around 1950. He wears a brass helmet with glass viewing ports. The helmet is connected to a brass collar worn on the shoulders, and is an integral part of the canvas and rubber suit covering the diver. Heavy lead weights are suspended over the chest and back, and lead-weighted boots overcome the diver's natural buoyancy, allowing him to walk upright on the ocean floor. An air hose and safety tether from a surface vessel complete his outfit.

Shown on the right is a more modern hard hat diving system. The diver wears a watertight suit of rubbery, foamed neoprene that is often electrically heated to ward off the cold of working in deep water. Cables from a surface support ship provide the air supply, communications, safety line, and electrical power. A weighted belt helps to stabilize the diver, and insulated rubber boots allow him to maneuver more freely around underwater installations such as oil-drilling rigs. Several types of more rigid "armored" diving suits have also been developed to enable divers to work safely at much greater depths. This equipment is designed to protect the diver from the crushing water pressure encountered while working at depths of 500 to 3,000 feet.

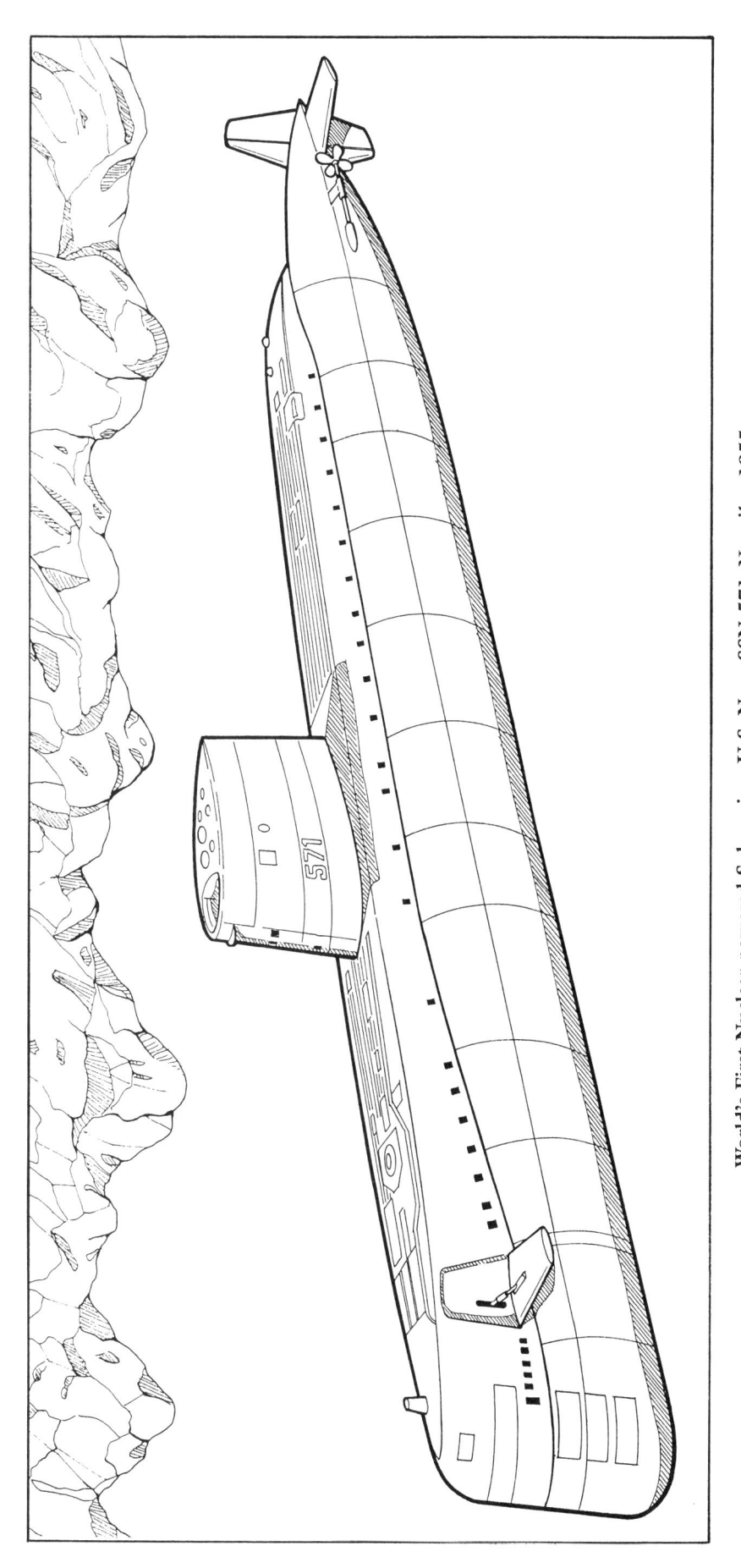

World's First Nuclear-powered Submarine, U.S. Navy SSN-571 *Nautilus*, 1955

The nuclear-powered submarine *Nautilus*, launched in 1955, was the final culmination of underwater vessels bearing that name. Beginning with Fulton's primitive *Nautilus* of 1801, and inspired by Jules Verne's fabulous fictional *Nautilus*, the U.S. Navy's *Nautilus* was the first ocean-going vessel of any kind to be propelled by the energy of atomic fission. Designed and built under the direct supervision of Admiral Hyman G. Rickover, the *Nautilus* was the realization of Captain Nemo's submarine boat, able to travel submerged over great distances and at high speed.

The heart of this historic submarine was the nuclear fission reactor powered by an almost inexhaustible supply of uranium. Without the need to replenish diesel fuel or recharge electric batteries, the *Nautilus* could cruise unrefueled for many months and thousands of miles. The propulsion system of the *Nautilus* utilized the heat from the radioactive decay of her fuel to boil water and create steam. Under high pressure, the steam is then used to spin turbine blades that turn the propellers. Her designer, Admiral Rickover, is honored as the "father of the American nuclear navy," now comprising dozens of submarines and surface warships. Rickover, considered a genius of naval engineering, overcame enormous technical and bureaucratic difficulties in launching the *Nautilus* by the sheer force of his personality.

This remarkable boat could cruise on the surface at a rapid speed of eighteen knots, but more remarkably, could travel at twenty-three knots submerged. She was 323-feet long, had a twenty-eight foot beam, and displaced 3,674 tons. Her armament consisted of six torpedo tubes. In 1958, the *Nautilus* became the first submarine to travel completely under the North Polar ice cap. Her record-breaking voyage began in Pearl Harbor, Hawaii, and ended in Portland, England, via the North Pole. She traveled an amazing 62,652 miles on her first nuclear core, 91,324 miles on her second nuclear fuel supply, and a stunning 150,000 miles on her third nuclear fuel supply. The *Nautilus* was retired from service in 1979 and is now on display in an historical museum at Groton, Connecticut.

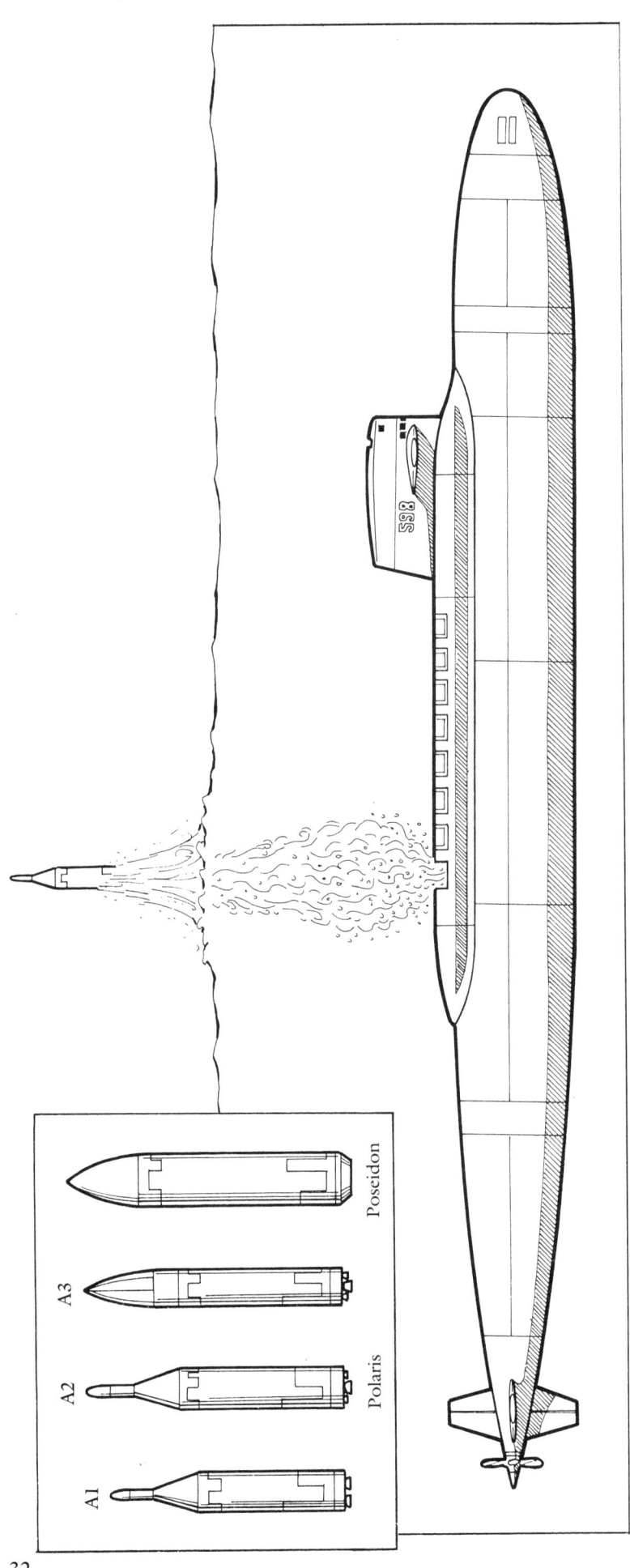

American Fleet Ballistic-Missile Submarine (SSBN), *George Washington* Class, 1959

Submarines underwent a period of intense and aggressive technological development beginning in the 1950s, fueled by the "Cold War" between the democratic nations of the free world and the totalitarian states of the Communist empire. The principal adversaries in this undeclared war were the United States and the Soviet Union. The Cold War lasted from 1948, when the Soviets blockaded the free city of West Berlin, to 1991, when the Soviet Union finally collapsed under the weight of its own heartless bureaucracy. During this era, the two nations competed for military supremacy by building nuclear arsenals of tremendous destructive capability. Their combined stockpile of nuclear warheads contained enough devastating power to completely annihilate each other as well as most of the rest of the world.

The American strategy of nuclear war deterrence was called Mutually Assured Destruction (MAD). It was based on the premise that any nation attempting a first strike against the United States with atomic weapons would be bombarded with sufficient retaliatory nuclear missile strikes to completely destroy the aggressor. This policy kept an uneasy peace and prevented an all-out nuclear war for nearly fifty years. To implement MAD, the American strategy depended on a so-called "nuclear triad" which consisted of long-range manned strategic bombers carrying nuclear bombs; land-based intercontinental ballistic missiles with multiple nuclear warheads; and submarine-launched nuclear-armed ballistic missiles. To this end, the world's first nuclear-powered fleet ballistic-missile submarine, the *George Washington*, was launched by the U.S. Navy in 1959. She was the first of many such fast and deadly missile boats.

The *George Washington* was armed with sixteen Polaris nuclear-tipped missiles designed to be fired while the boat was submerged. The first model of the Polaris, the A-1, had a range of 1,300 miles. Subsequent models had their range increased to 2,500 miles. She was also equipped with six conventional torpedo tubes. The boat was 381-feet long, and displaced 6,019 tons. Designed with a revolutionary cylindrical teardrop shape, she had a surface speed of twenty knots, and a very fast submerged speed of thirty knots. A total of forty-one boats in the *George Washington* class and its successor—the *Lafayette* class—were built during the 1960s. In 1970, the U.S. Navy started converting most of the Polaris submarines to fire the improved Poseidon missile.

32

Submersible Bathyscaphe *Trieste* Descends to Record Depth, 1960

On January 23, 1960, researchers Jacques Piccard and Lieutenant Don Walsh of the U.S. Navy made a world record descent to 35,817 feet in the *Trieste*—a deep-diving submersible research vessel. The bathyscaphe descended to the sea floor via the Challenger Deep—a 36,000-foot hole discovered by the British oceanographic vessel Challenger II—part of the Mariana Trench located east of the Phillipines. There the passengers witnessed life in the form of a flat fish and a few shrimp, alerting scientists to the hitherto unknown fact that fish could live even in the deepest ocean depths. The *Trieste*'s momentous trip to the bottom of the sea took nearly five hours each way.

The *Trieste* is a "bathyscaphe" type of submersible which was constructed in two parts. The upper portion consists primarily of a large cylindrical tank filled with 40,000 gallons of gasoline. Because gasoline is lighter than seawater, it provides the vessel with flotational buoyancy. Additionally, since gasoline cannot be compressed, it helps prevent the *Trieste* from being crushed by the immense water pressure at its operational depth. The lower part of the *Trieste* contains a steel sphere whose walls are four inches thick, around six feet in diameter, and is used to carry the underwater explorers. On either side of the passenger sphere are hoppers containing iron pellets that were used as ballast to help the craft descend, and then emptied to help it rise back to the surface. Mounted on the exterior are searchlights, microphones, other test equipment, and a propeller for additional aid in descent and ascent.

Submersible Research Vessel, *Alvin*, 1964

The *Alvin* deep-sea research vessel shown above is one of the most versatile submersibles ever built. It has been operated by the Woods Hole Oceanographic Institute since 1964. *Alvin* can carry a pilot and two passengers within its spherical titanium pressure hull to a depth of 13,000 feet. It is equipped with a remote manipulator arm and external basket for collecting samples from the ocean floor.

In 1966, *Alvin* was used to locate a sunken nuclear bomb lost after two American bombers collided over the sea near the coast of Spain. The nuclear device was retrieved from a depth of 2,850 feet. Two years later, while *Alvin* was involved in another salvage operation, it accidentally sank in 5,500 feet of water. No one was aboard at the time, and it was rescued by another deep-diving submersible, the *Aluminaut*. The *Alvin* was completely rebuilt after its sinking, and continued to engage in underwater research.

In 1977, *Alvin* and its crew made an important discovery. They found "hydrothermal vents"—jets of hot water shooting up from cracks in the seabed at a depth of 12,500 feet. The first of these vents was found near the Galapagos Islands, off the South American coast. Since then, other vents have been located and explored. These hot water streams are caused by heat from the earth's molten interior. As the water shoots out of the vent, sulfur and other minerals are deposited around the edges. Over time, this sediment builds vertical columns or "chimneys" of solidified minerals around the vents which can reach a height of thirty feet. Because the water from these vents is colored black from the sulfur content, these chimneys have been named "black smokers." The chemical composition and excessive heat of the water (550°F) around the black smokers makes the area poisonous to most marine life. However, researchers have found a few specialized life forms that seem to thrive in this environment. Shown above are "tube worms," creatures that can grow to a length of ten feet. They have neither mouths nor stomachs but live off the internal bacteria found in their bodies. Other unusual marine animals found around black smokers are crabs and lobsters without eyes, and clams and mussels that can grow up to one foot in length.

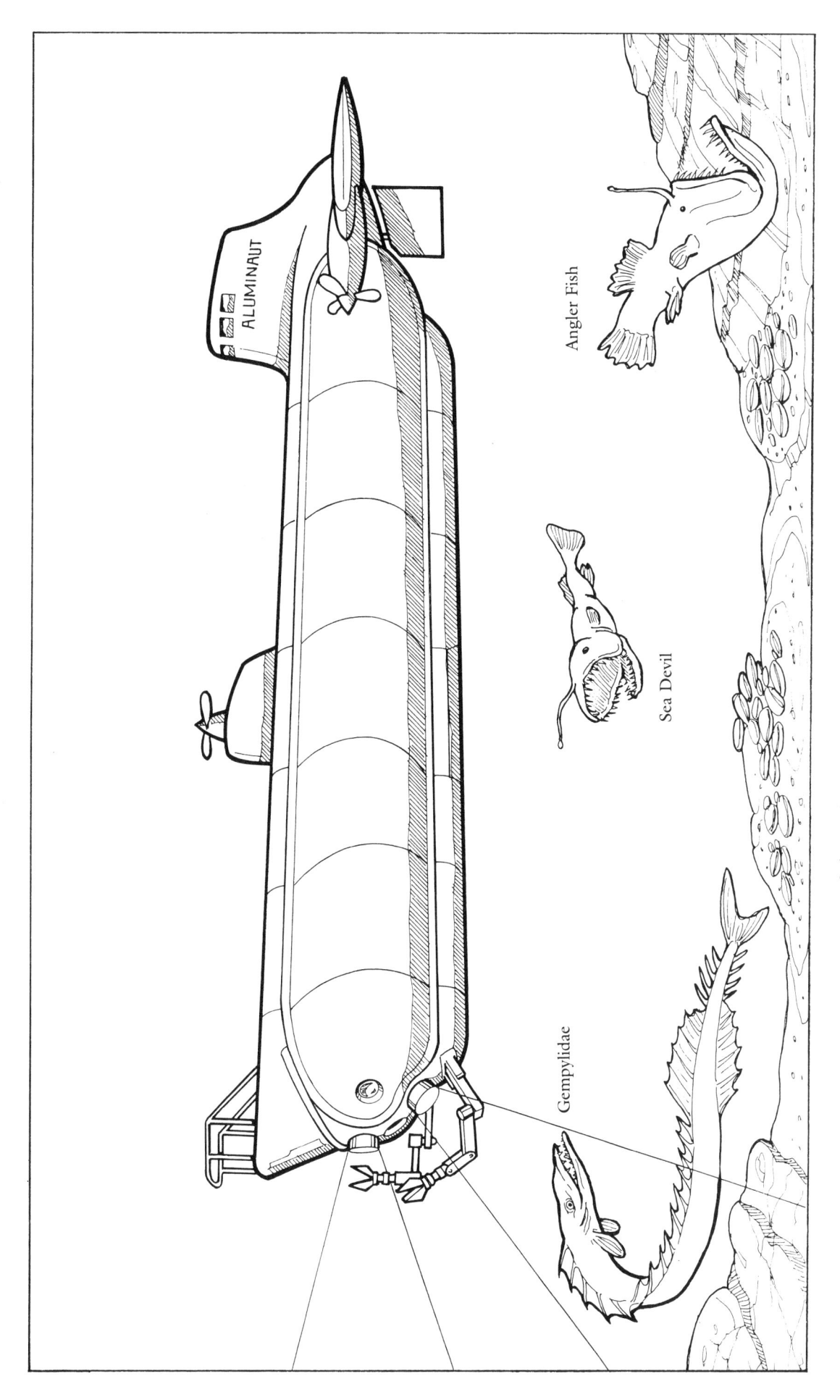

Submersible Research Vessel, *Aluminaut*, 1968

One of the largest research submersibles to explore the deep ocean was the sixty-five foot long *Aluminaut*. Its inner pressure hull had six-inch thick walls fabricated from aluminum. The *Aluminaut* could carry six passengers to a depth of 15,000 feet. It was equipped with grappling arms and a top-mounted propeller for vertical maneuvers. To aid in its descent and ascent, *Aluminaut* also carried both seawater and iron pellets as ballast.

The *Aluminaut* was the primary rescue vessel in the successful salvage of the submersible *Alvin* when, in 1968, the smaller vessel accidentally sank in 5,500 feet of water. *Aluminaut* played a crucial role in *Alvin*'s location and recovery.

1986 Soviet *Akula* Class

1976 American *Los Angeles* Class

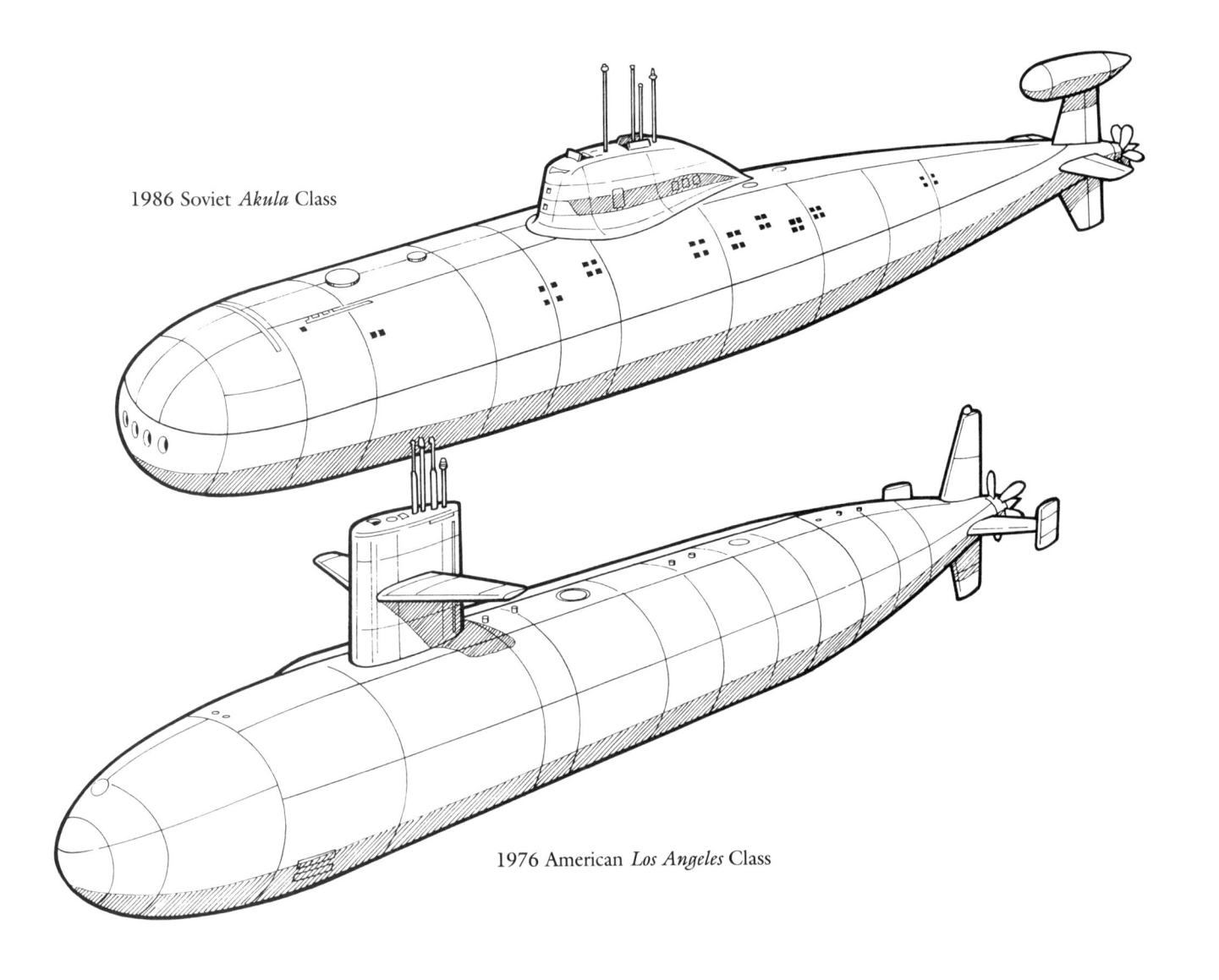

Soviet *Akula* Class Attack Submarine, 1986 and American *Los Angeles* Class Attack Submarine (SSN), 1976

As the Cold War continued during the 1970s and 1980s, the United States and the Soviet Union maintained large fleets of advanced submarines. These included both nuclear missile firing boats and hunter-killer attack submarines primarily designed to find and destroy the missile boats.

Shown above are examples of a Soviet *Akula* class attack boat and its adversary, an American *Los Angeles* class attack sub. The *Akula* class boats were introduced in 1984 and were an outgrowth of the earlier *Victor* and *Alfa* classes. These deep-diving boats could reach an operational depth of 1,300 feet, and a record-breaking underwater speed of thirty-five knots. Their streamlined hulls were built from titanium and featured a teardrop-shaped sonar array mounted on the stern fin. (SONAR is a sound-echoing system used to detect and locate underwater objects. It is the principle means used by submarines to track other underwater vessels.) Boats of the *Akula* class are 370-feet long, have a 42-foot beam, and displace 7,500 tons. They are fitted with six forward launch tubes that can fire anti-ship or anti-submarine torpedoes, as well as anti-ship missiles.

The American *Los Angeles* class boats began operational service in 1976. They were developed from the earlier *Permit* and *Sturgeon* class nuclear-powered attack subs. For a hunter-killer submarine, quiet operation is the key element for successful undersea warfare. The *Los Angeles* class boats were designed for extremely quiet operation to avoid detection by enemy ships and submarines. Their 35,000 horsepower nuclear-powered steam turbines can propel the boats at swift underwater speeds in excess of thirty knots. They are also capable of operating at a depth of 1,425 feet. The *Los Angeles* class submarines are 360-feet long, have a beam of thirty-three feet, and displace 6,080 tons. One of the most impressive features of these attack vessels is their variety of weapons. They can fire conventional and wire-guided torpedoes, Subroc and Subharpoon anti-ship missiles, and Tomahawk cruise missiles from their various launch tubes. This class of submarines forms the majority of the American navy's attack boat force with a total of sixty-six vessels currently in service.

Remotely Operated Vehicle (ROV) *Jason Jr.* and *Alvin* Explore Wreck of the *Titanic*, 1986

The story of the *Titanic*, the doomed ocean liner lost at sea in 1912, has long fascinated the world. The great ship sank in 12,500 feet of water after striking an iceberg in the north Atlanic. The search for the wreckage of the *Titanic* has also drawn the attention of many underwater explorers. The sunken ship was finally discovered and located in 1985 by a towed submersible vessel, the *Argo*. In 1986, the wreckage was explored and photographed by the veteran research submersible *Alvin* and its smaller partner, *Jason Jr.*

Jason Jr. is one of a new class of underwater exploration vessels developed during the 1970s called Remotely Operated Vehicles, or ROVs. They are essentially robot submersibles operated from distant locations. *Jason Jr.* was operated from *Alvin* and connected by a 280-foot electrical cord. The small robot vessel is only twenty inches by twenty-seven inches by twenty-eight inches. Its on-board thruster motors could maneuver the mini-submersible into fairly tight spaces. In 1986, *Alvin* and *Jason Jr.* explored the wreck of the *Titanic* at a depth of 12,500 feet. The project was under the direction of Dr. Robert Ballard, head of the Deep Submergence Lab and one of the world's foremost underwater research scientists. Dr. Ballard and his team maneuvered the diminutive robot vehicle into the hulk of the *Titanic*, and through various corridors and rooms. Spectacular video and still photos taken by *Jason Jr.* showed the very well-preserved condition of the ship's artifacts. Since 1986, the *Titanic* wreckage site has been visited numerous times by *Jason Jr.* and other deep-sea exploration vessels.

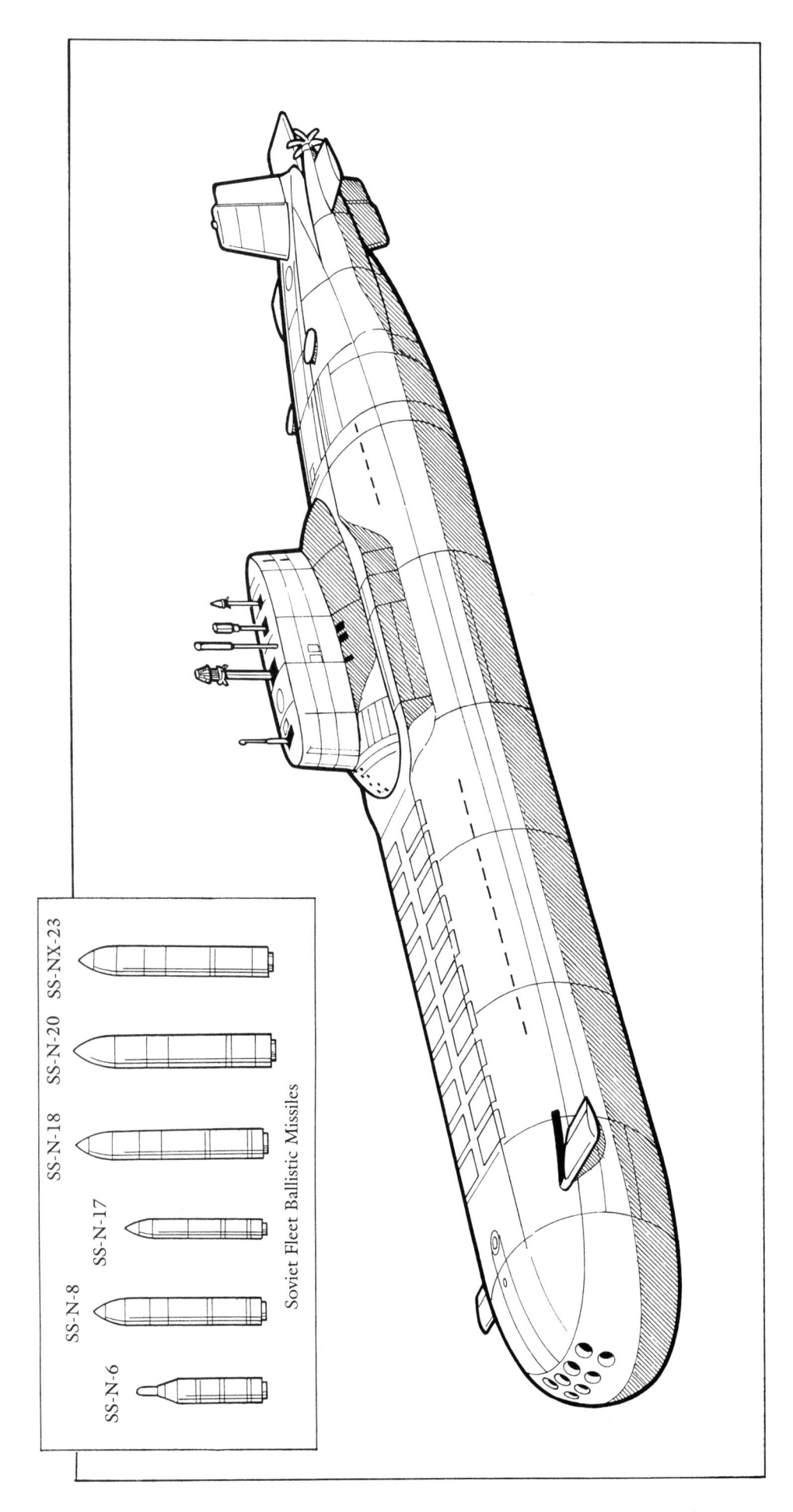

Soviet *Typhoon* Class Ballistic-Missile Submarine, 1983

The keystone of the Soviet nuclear-powered ballistic-missile submarine fleet is the enormous *Typhoon* class of submarines, which at 561-feet long are the size of a heavy cruiser of World War II vintage. With a very broad beam of seventy-nine feet, these immense submarines displace a whopping 18,500 tons, and are the largest underwater vessels ever constructed.

The *Typhoon* class evolved from earlier versions of large Soviet missile boats of the *Delta* and *Yankee* class. The principal armament of the *Typhoon* boats is a bank of twenty missile tubes mounted forward of the conning tower (also called the sail). These contain the extremely powerful SS-N-20 intercontinental ballistic missiles that have a range of 4,300 miles, with each missile containing 6–9 MIRV (Multiple Independent Reentry Vehicles) nuclear warheads of tremendous destructive power which can be targeted at different locations. American missile submarines as well as land-based intercontinental ballistic missiles also contain MIRV warheads.

The *Typhoon* subs are also equipped with torpedo tubes that can fire anti-ship missiles or conventional torpedoes. These massive boats are powered by two nuclear reactors with a combined rating at 75,000 horsepower. Their twin propellers can drive them at a submerged speed of twenty-five knots. The maximum depth capability of the *Typhoon* class is unknown but is estimated at over 1,200 feet.

Soviet Fleet Ballistic Missiles

SS-N-6 SS-N-8 SS-N-17 SS-N-18 SS-N-20 SS-NX-23

American Fleet Ballistic Missiles

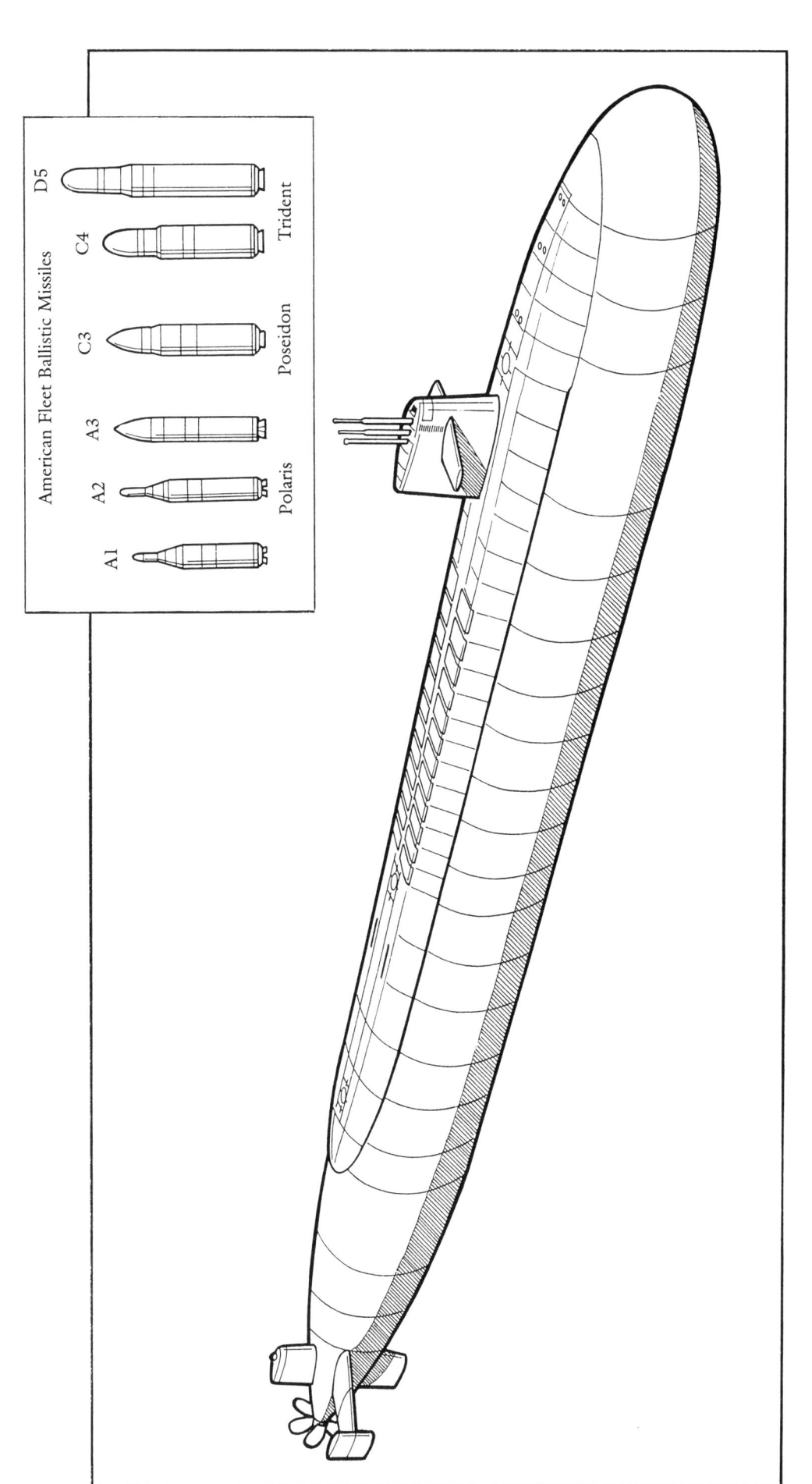

- A1
- A2 Polaris
- A3
- C3 Poseidon
- C4
- D5 Trident

American Fleet *Ohio* Class Ballistic-Missile Submarine, 1981

The heart of the U.S. Navy's ballistic-missile submarine force is the very advanced *Ohio* class nuclear-powered warship. Introduced in 1981, these boats are the quietest submarines in operational service with the exception of the *Seawolf* class of American attack subs just entering service in 1996. The *Ohio* boats are the successors to an earlier missile boat series, the *Lafayette* class, launched during the 1960s. The *Lafayette* subs were designed to fire the Poseidon ballistic missile, the more advanced version of the Polaris. The *Ohio* class submarines are armed with the even more sophisticated Trident missile, which has a range of 4,200 miles and can carry eight MIRV nuclear warheads. These are fitted in two rows of twelve launch tubes located aft of the sail. The boats are also equipped with four torpedo tubes for firing conventional underwater torpedoes. With its tremendous destructive firepower, an *Ohio* missile boat with Trident missiles can be positioned to strike any target on earth without prior detection. *Ohio* missile submarines are almost as large as the Soviet *Typhoon* class. They are 560-feet long, have a beam of forty-two feet, and displace 16,600 tons. Their single nuclear reactor drives twin steam turbine engines rated at 60,000 horsepower. They can travel at over twenty knots submerged and can reach a depth of 1,000 feet.

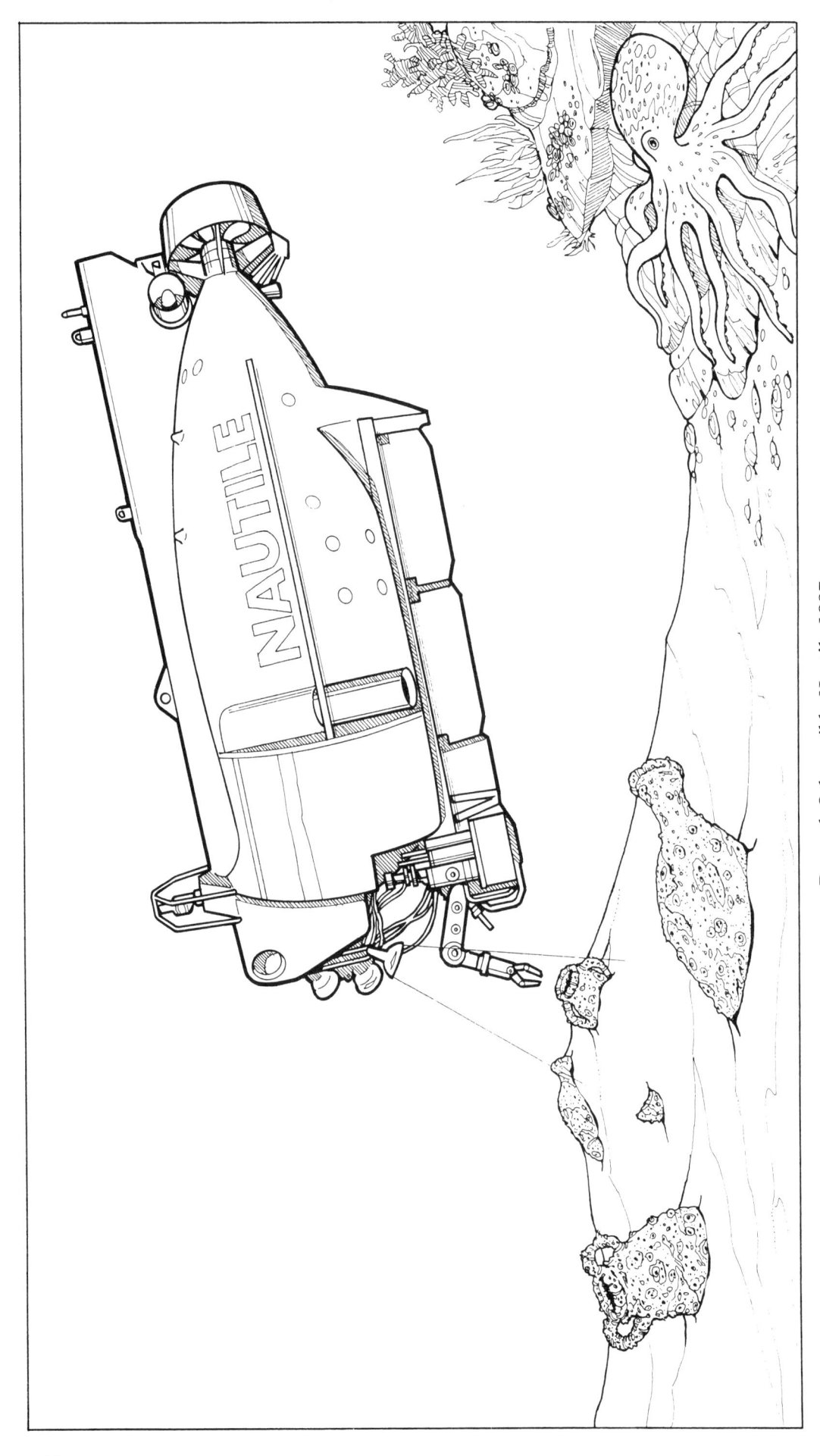

Research Submersible *Nautile*, 1987

One of the current class of underwater research vessels is the bright yellow submersible, *Nautile*, one of the deep-diving submersibles used to explore the ocean floor around the wreck of the great ocean liner *Titanic*. When it sank in 12,500 feet of water, the hull broke into two pieces, scattering wreckage over a wide area of the sea bottom. The *Nautile* was able to reach and explore this debris field, recovering numerous objects with its remotely operated manipulator arm.

Within the twenty-six foot long aluminum hull of the *Nautile* is the pressure-resistant compartment for its occupants. This water-tight sphere is fabricated from titanium, a metal stronger than steel but lighter than aluminum. Such a tremendously sturdy material is needed to withstand the great pressure of the ocean water at the *Nautile's* operating depths. The vessel is equipped with thick, curved Plexiglas portholes that actually flatten out from water pressure as the *Nautile* descends toward the sea bottom.

The *Nautile's* titanium pressure sphere can carry three passengers: a pilot, a co-pilot, and an observer—usually a marine scientist or underwater archaeologist. The submersible has an air supply sufficient for eight hours of deep-sea exploration.

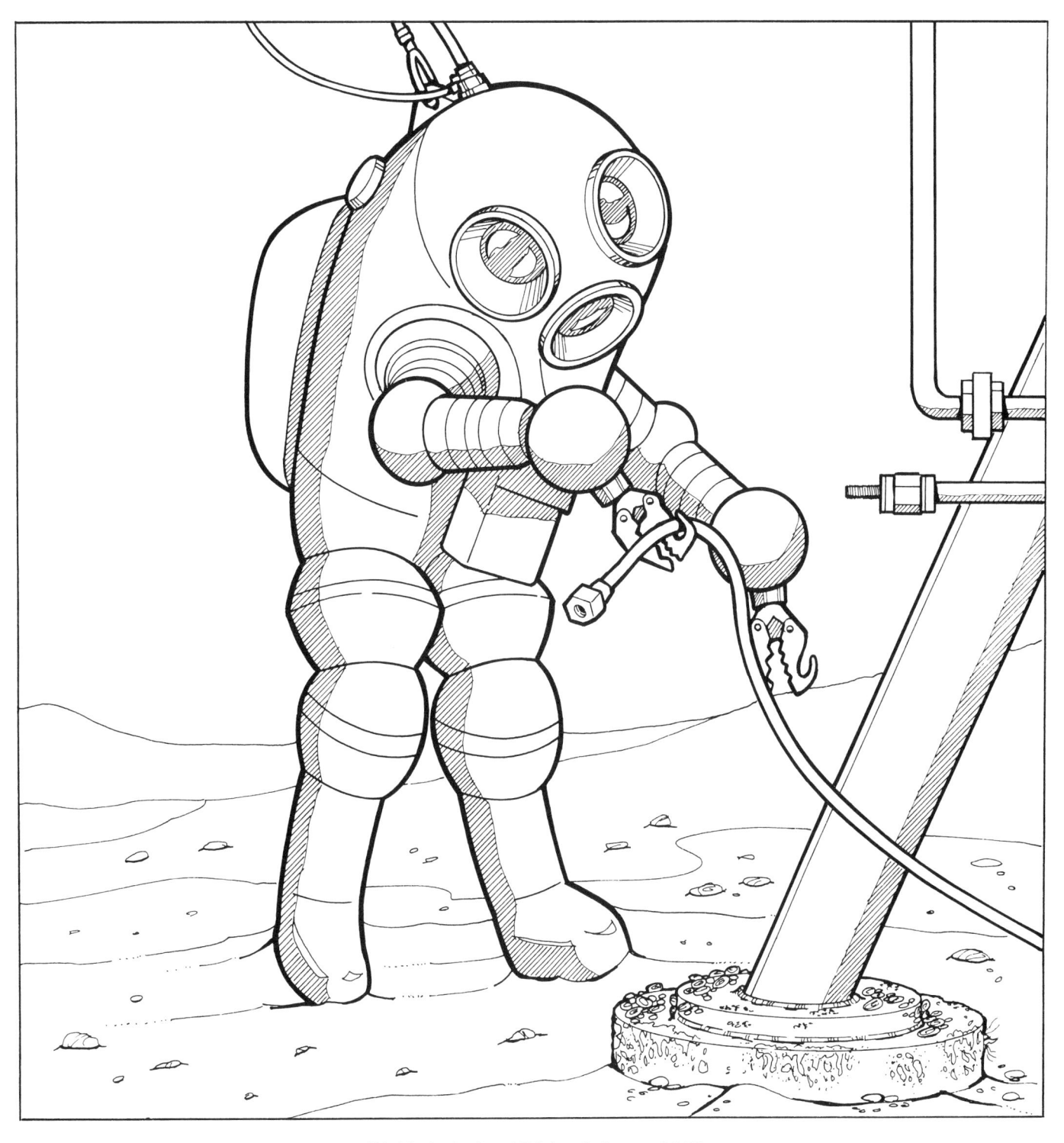

Rigid, Articulated Diving Suit, JIM, 1975

Shown above is an example of a rigid, articulated diving suit used for underwater activity at depths exceeding the capabilities of SCUBA or conventional hard-hat diving equipment. Those systems are limited to around 200 feet and 500 feet, respectively. The type illustrated is called a "JIM" suit, after diver and mechanic Jim Jarrett who worked on its development and testing. The JIM suit can operate at depths approaching 1,500 feet. At those depths, the water pressure approaches 800 pounds per square inch. JIM's light but strong armor of magnesium alloy metal protects the diver and keeps the pressure inside the suit at the normal surface atmospheric pressure of fourteen pounds per square inch. Because of this, diving suits like the JIM are called "Atmospheric Diving Suits" (ADS).

The occupant of a JIM suit wears a face-mask and breathes from an air tank mounted on the back of the suit. A device called a "scrubber" removes the poisonous carbon dioxide gas exhaled by the diver. Although the suit weighs 1,000 pounds and looks quite cumbersome, the diver has considerable flexibility by virtue of the oil-filled ball and socket joints for the arms and legs. The grabber claws at the ends of the arms are manipulated by the divers hands from within the suit. The suit is lowered and raised by a steel cable connected to a surface ship. The diver's electric power and communications lines are also cable-connected. The SPIDER and the NEWSUIT are examples of other atmospheric diving suits designed to protect divers from the cold and pressure of working at extreme underwater depths.

Modern SCUBA Diver

Underwater exploration has grown into a worldwide recreational activity since the development of SCUBA diving equipment in the 1940s. Thousands of amateur diving enthusiasts are able to swim and observe the undersea world thanks to the availability and flexibility of scuba gear.

Shown above is a diver equipped with current scuba equipment. She wears a neoprene rubber "wet suit" for warmth, which traps water in a layer next to the skin. This trapped water layer is warmed by the body and insulates the diver. The single air tank on her back allows for around forty minutes of underwater activity. The air demand valve in the mouthpiece regulates air intake and carbon dioxide expulsion. A weight belt worn around the waist counteracts the diver's natural bouyancy. She is seen next to a manta ray, a sea creature that swims by flapping its wing-like fins which can reach a substantial width of twenty-three feet.

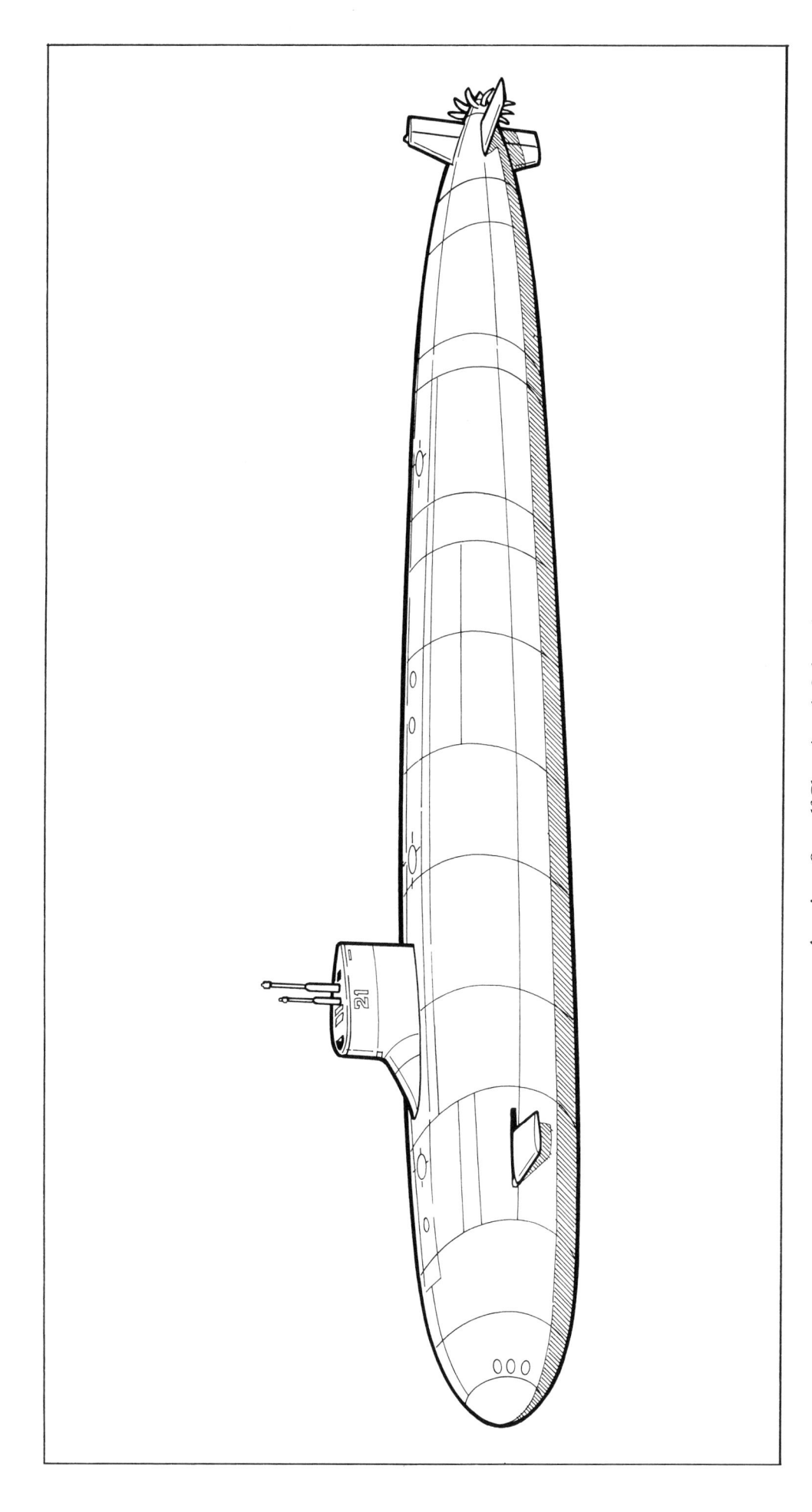

American *Seawolf* Class Attack Submarine, 1996

The latest addition to the American Navy submarine fleet is the nuclear-powered attack boat, SSN-21 *Seawolf*. The most advanced undersea vessel to date, it entered into operational service in 1996, the first of three boats in this class to do so. The second vessel is currently nearing completion and has been christened the *Jimmy Carter*. It has been named after the former president, a naval academy graduate who served as a submarine officer under Admiral Rickover.

The *Seawolf* is the quietest submarine ever developed. At a submerged speed of twenty-five knots, it is more noiseless than the already super-silent *Los Angeles* class boats. It is also very fast and a deep diver. Although exact figures for those performance statistics are top secret, it is estimated to have an underwater speed in excess of forty knots and a depth capability in the 2,000-foot range. The *Seawolf* is powered by a single nuclear reactor driving two steam turbines that deliver 52,000 horsepower.

The boat is 353-feet long and has a forty-foot beam. It displaces 9,150 tons submerged, and has a crew of 130 sailors and officers. The *Seawolf* is heavily armed with the latest submarine weapons. It carries fifty-two torpedoes, can launch Harpoon anti-ship missiles, mines and decoys, and Tomahawk cruise missiles. The *Seawolf* class boats will lead the U.S. Navy's submarine force well into the twenty-first century.

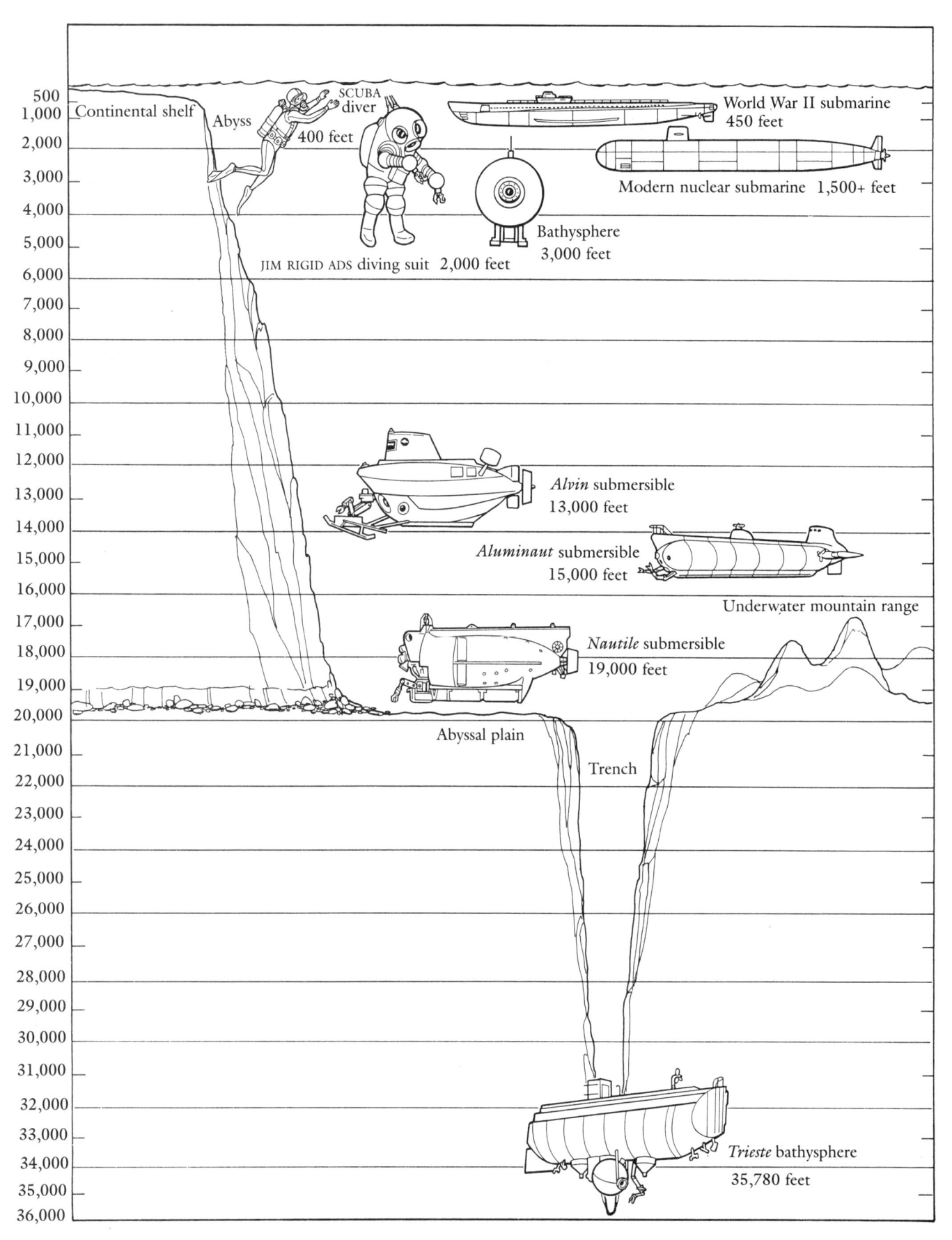

Data for Diving Apparatus—Depth Comparison Chart

Markings for depth are in 1,000-foot increments starting at the surface (0) and ending at 36,000 feet at trench bottom. The first 1,000 feet is also subdivided with 500-foot hash marks.